# 26 Shades of Shale

Jeff Scott

methanol
press

***"Change the title and the leading lady. Keep the dialogue. The ideas wash off."***

MICHAEL DONAGHY

**Thank you:**

Billy

Cas

Chris

Graham

Michael

Rachael

Vicky

Vy

First published in Great Britain by
Methanol Press
2 Tidy Street Brighton East Sussex BN1 4EL

For a complete catalogue of current and forthcoming publications please write to the address above, or visit our website at www.methanolpress.com



ISBN 978-0-9568618-2-5
A catalogue for this book is available from the British Library

Editor in absentia: Michael Payne
Word Wrangler: Graham Russel
Stenographer: Vy Shepherd
Book Design: Vicky Holtham
Cover Design: Rachael Adams
Cover Photograph: Jeff Scott

Printed in the UK by Imprint Digital

# Contents


| | | |
|---|---|---|
| Chapter 1 | Mildenhall v King's Lynn<br>**"If people want to say nice things that's good – but you shouldn't going looking for it"** | 6 |
| Chapter 2 | Ipswich v Plymouth<br>**"In fairness he was lucky to get that – I'd have given him minus one!"** | 12 |
| Chapter 3 | Rye House v Somerset<br>**"They've not got their ginger-haired Australian"** | 19 |
| Chapter 4 | Leicester v Berwick<br>**"It's like a Redcar setup but better!"** | 26 |
| Chapter 5 | King's Lynn v Coventry<br>**"Looks a little grip. I must make good track. Good race!"** | 34 |
| Chapter 6 | Somerset v Leicester<br>**"I thought they'd travel today but there's none here!"** | 41 |
| Chapter 7 | Edinburgh v Sheffield<br>**"He'll come out of the gate like a salmon and then he'll come out of the bend like a whale"** | 46 |
| Chapter 8 | Berwick v Ipswich<br>**"There used to be one bloke in a Ford Cortina with a bike on the back and he did everything! Now they have two bikes and a mechanic, at least!"** | 50 |
| Chapter 9 | Glasgow v Ipswich<br>**"In those days, we didn't talk to the riders"** | 57 |
| Chapter 10 | Birmingham<br>**"'Make Me A Fan' scheme. It's called promoting!"** | 60 |
| Chapter 11 | Newport v Dudley<br>**"If speedway was raced over one lap, James White-Williams would win all the races!"** | 68 |
| Chapter 12 | Cardiff: SGP<br>**"We tried to get away from the Wulfsport look, so we designed our own!"** | 75 |
| Chapter 13 | Wolverhampton v Coventry<br>**"They're a nice couple – those Ludvigs!"** | 81 |
| Chapter 14 | Poole: U21s<br>**"We all know the riders will turn up and ride for the goblin's gold"** | 88 |
| Chapter 15 | Dudley: Golden Hammer<br>**"Imagine if he was to come back wearing the Hammer and Heathen on his chest?"** | 96 |

Chapter 16 | Redcar v Workington | 103
**"He'd fall off so he could be first in the bar"**

Chapter 17 | Scunthorpe v Glasgow | 110
**"When the dirt builds up on that bend two and the riders go out wide, they can almost disappear in it up to their handlebars!"**

Chapter 18 | Workington v Glasgow | 116
**"I'd take two broken legs over this!"**

Chapter 19 | Glasgow v Edinburgh | 122
**"He may be a smidgen over 17 but he rides like he's 35!"**

Chapter 20 | Belle Vue v Coventry | 128
**"He replaces the other Pole who rode in Poland when he was injured!"**

Chapter 21 | Lakeside v Swindon | 136
**"Is that your sixth times World Champion ring?"**

Chapter 22 | Swindon v Peterborough | 144
**"Tyres take ten minutes to fit and silencers two minutes!"**

Chapter 23 | Sittingbourne: U15s | 151
**"I set off and something went whoosh and I thought, what's that?"**

Chapter 24 | Newcastle v Glasgow | 156
**"We need more ideas and more young people. What speedway also needs is more fights!"**

Chapter 25 | Isle of Wight v Scunthorpe/Sheffield | 162
**"I watch from the balcony but I can't watch the first corner"**

Chapter 26 | Sheffield v Workington | 171
**"He's been doing that all season – turning left!"**

Afterword | 178

CHAPTER 1

# Mildenhall

## "If people want to say nice things that's good – but you shouldn't going looking for it"

15th May

Rider parade

The mid-season demise of the Mildenhall Tigers and their sudden exit from the 2010 National League bothered speedway fans generally in addition to their supporters. After some uncertainty, news of a beauty contest between would-be speedway promoters conducted by stadium owner Dave Coventry indicated bikes would, once again, roar at West Row stadium. However, the steady stream of promoters over recent years plus an ongoing noise complaint from a local resident would create uncertainty anywhere. Prospective speedway promotional joint partnership of Kevin Jolly and Steve Ribbons saw their application to rejoin the National League narrowly rejected at the NL AGM held in Rugby (on 25th January in a 4-3 verdict from existing voting members). Understandably, Steve Ribbons reacted with incredulity, "I'd a licence at Rye House and Wimbledon, and never had it rescinded, suspended or revoked, but at Mildenhall I wouldn't have been allowed into my own pits area. It was a clear indication to me that the list of conditions outlined to me over the telephone by [Peter] Morrish were a personal issue. I was asked for a £5,000 payment for League membership, another £5,000 for the 'promotional rights', plus a bond of £3,500. Then I was told I wouldn't be invited to the NL meeting at which my bid was being considered. The way it operated was like a judge giving a verdict on the first day, without hearing any evidence. Kevin Jolly was invited to the meeting, but only to be told the results of the vote." Letting his heart rule his head, Steve Ribbons took to the British Speedway Forum to range widely as well as address various critiques of his bid raised by other posters.[1]

[1] "For those who wonder why I am at loggerheadswith the BSPA its down to one man.... (I don't believe he could co-ordinate his shirt with his underpants. He is a fool and I have told him he is a fool! the very first time I telephone him he berated me for trying to start Rye House and we were on a downer all that season with him once saying to me (in front of witnesses "you lot are just gipsy's, you'll be selling heather next!)... Lets examine his qualifications shall we.... I wouldn't trust him to run me a bath let along a league!!! Is it wrong to Post this?...yes, but I am sick and tired of thse personal attacks on me because of the fact I deeply care about the sport ofspeedway racing. I can say all this because I've given up trying to be polite to those people who don't like me behind my back but aren't man enough to say it to my face. Good luck to who ever takes over at Mildenhall.... But for those of you who do try to get off our asses and try to do something for the good of the sport and open yourself up to public ridicule, my advice is...."

Elsewhere National League co-ordinator Peter Morrish said, "It's a great little track, and there's no better place to watch speedway. There's no doubt we need clubs such as Mildenhall in the sport, especially as it is a stand-alone track. But there were concerns about various aspects of this bid, which is why the conditions, which did not seem unreasonable, were set." Fine words usually butter no parsnips so many Fen Tigers fans resigned themselves to a 2011 season of NL exile. By March, the High Court ruled the noise nuisance case against the West Row stadium owners should be reassessed (in October) after further summer noise level monitoring. Since West Row stadium is under the Mildenhall airbase flight path, fans took reassurance from the lack of restrictions or curfews imposed.

Proving that a week is a long time in speedway and politics, ex-Mildenhall rider Kevin Jolly quickly emerged (phoenix-like) from the flames as part of a promotional triumvirate with two ex-England International riders, Michael Lee and Chris Louis. Only days after the initial NL AGM disappointment, Chris Louis contacted Kevin to attend a meeting to thrash out prospective plans with stadium owner Dave Coventry and his brother, Ron. Chris Louis explains, "I talked to Michael who was as keen as me on the idea of a Speedway Academy and things just developed from there." The triumvirate quickly gained agreement with the stadium and also got green lighted by speedway authorities keen to offer the new Mildenhall promotion "all the support they can to rebuild the club". An elated Kevin Jolly commented, "It's great news for Mildenhall and our loyal supporters and I believe also for East Anglian and British speedway. Because a huge part of our plan is to set up a Speedway Academy with regular training sessions at West Row run by myself and two of the best riders in the history of speedway, Michael and Chris." Subsequently Michael Lee said, "British clubs will never be able to offer all the opportunities that Poles are given now, with academies in place at every track. I look at them and can see possibly eight youngsters, some I've never even heard of, who could go on to become world champions. It's difficult, but we've got to do something to help our own lads and that's what we want to be about at Mildenhall."

If young rider development is the cornerstone of the promotional triumvirate's strategic plan, actually running meetings is the bread and butter activity of any club. With early season crowds everywhere boosted by unseasonably warm dry weather, the Fen Tigers first home meeting saw a bumper crowd celebrate the return of their club (also boosted by notable Dudley away support). Since then the Tigers beat Belle Vue and, for their third home meeting, enjoy a local derby contest versus King's Lynn Young Stars. The rejuvenation of the speedway also fires the corporate imagination in Mildenhall High Town, if judged by the programme, "James and Lewis had a great time helping to open Wilkinsons new store in St Andrew's Street, Mildenhall. Big thank you to Wilkinsons who presented James and Lewis with a cheque for £250 for our new academy."

Under overcast skies, the tractor circles grading the circuit. The shale is a lovely dark colour but curatorial work is difficult since the surface dries quickly in the cool breeze. In the pits, well-travelled speedway fan Brian Oldham asks, "Are you still around then?" Stood by the stadium Armco wall, John Louis breaks off his conversation, "Oh no! He's here with his dodgy books!" Over at the trackshop by the grandstand Fenman bar, Johnny Barber patiently sets out the merchandise with skill under the watchful attention of Tigers uber-fan Michael Slattery, "Wait a minute, Michael. Let me put out the new mouse mats!" Mildenhall Fen Tigers mouse mats excite Michael, "When did they get done?"

Technology is central to modern life including Johnny's. He'll text heat by heat scores to the popular Speedway Updates website. In the programme, the anonymous Clippy writes, "I don't know if many of you read the British Speedway Forum, but in the Clippy household it is a regular source of amusement. The recent Ipswich/Rye House threads made me smile in particular as they degenerated into an 'our toilets are better than yours debate!' Well let me tell you the lavatory in the Clippy house is better than either, but then again we didn't have over a 1000 people using it on Bank Holiday Monday!" Ahead of the recent speedway Twitter trend, Johnny keeps tweets to a minimum. Sadly many others share endless platitudes about the minutiae of their lives. Johnny is much more *au fait* with Facebook, "You know you should go on Facebook and do a book based on the riders' Facebook statuses! It would be a bestseller. They're a book in themselves. Jerran Hart's are fantastic! They're full of text speak – or bang, bra and other stuff." The lure of Twitter is irresistible to some who seek speedway celebrity. "Every week Mike Bennett tells the crowd at King's Lynn 'I'm on Twitter if you want to tweet any requests'. When ordinary fans do, he says nothing but if Taylor Poole gets in touch he says 'um – I've got a tweet here from Taylor Poole!' Kevin Moore came in the shop recently and said 'Is there anything to plug?' and I said, 'Only Jeff Scott's new book'. 'Does he have a new one then?' After I told him you did he said, 'Do you think there'll be anything about me in it?' I told him I didn't know but he said, 'If there is, there'll be trouble!'"

On the British Speedway Forum, Mike Bennett kindly starts a thread and volunteers to visit Leicester Speedway club to selflessly pass on (gratis) his extensive speedway commentary and presentational experience. Few doubt Mike's facility with the equipment or words. Mike's fluency could lead the uninitiated to underestimate the difficulty of performing at a live event. As is the BSF way, subsequent posters express scepticism about the sincerity or usefulness of said offer (one notes Mike has an ego so big it requires its own postcode). The newly minted promotional team at Leicester speedway could do much worse than take him up on his offer. Like many, Johnny savours the BSF exchanges, "For someone like you – who writes – that must be pure gold!"

Smartly attired in his trademark white jacket, Mildenhall Tigers Speedway Master of Ceremonies aka club presenter Kevin Long could pass as a maître d of a smart restaurant, given his formal attire set off by closely shaven head and chin. Kevin doesn't have much time for self-aggrandising presenters who he believes unnecessarily draw attention to themselves with *faux* modest comments on Facebook, Twitter and various other websites. "I'm on Twitter at Witches' request and fans can send in messages to be read out during the meeting via the Ipswich website. But I really believe that you should be true to yourself, do the best job you can and let your work speak for itself. If people want to say nice things that's good – but you shouldn't go looking for it. I'm one of Mike Bennett's followers on Twitter, I don't know why!"

With numerous people to meet and greet during his pre-meeting preparations, Kevin cuts to the chase, "How many books is it now?"
[Jeff] "Six."
[Kevin] "What's the name for that?"
[Jeff] "Dunno."
[Kevin] "Come on, you're the intelligent one".
[Jeff] "Dunno. Let's say it's a double trilogy then."

Kevin planned to go on the Ipswich Witches Supporters mini Northern tour (organised by Chris Ellis). Sadly, it's been fated from the outset. Initially billed as a three-leg tour to Scunthorpe (Friday), Workington (Saturday) and Newcastle (Sunday), the Brough Park leg quickly fell victim to dire circumstances. "Some greyhounds died because of contaminated sand so William Hill – the stadium owners – want it fixed immediately. So George English had no choice but to cancel the meeting. He appealed to them saying that we were on tour but it's that serious they had no choice." Despite this cancellation, the tour itinerary switched to include Leicester versus Plymouth or did right until two days ago when the Scunthorpe staging of the British semi-final fell foul of the weather and was re-arranged to seven days later. This, thereby, struck out the Scunthorpe Ipswich KO cup meeting from the schedules. Kevin isn't happy, "I'm very disappointed about the tour being cancelled. At Newcastle you can understand but with the dogs dropping like flies they had to cancel. I hadn't been there before or to Workington. The riders would have been along too. Staying in the hotel, it was a proper tour. I've been to Scunthorpe before but – why organise a meeting on a night, which is a rain-off for the British Final semi-final? Okay, they've got to have a cut off and they've got to have a rain-off date, but why organise our meeting when they know we're on a tour? It's not the result. We're 20 points up from the first leg – they would have been comfortable no matter what. I'm so annoyed I'm thinking of writing to the *Star* and I've never written to them before! I won't be going to the rearranged Scunthorpe meeting on 3rd June now – which is a one off – because it clashes with the Mildenhall horserace evening at Worthington Hall."[2] "I brought Chris Ellis with me – she's feeling better today [after cancelling the tour yesterday] – as Bob is doing the track. I help her a bit with Cardiff because she does so much to organise all the other trips. It's good to be a sounding board for her when things go wrong."

Kevin's diatribe is interrupted by the arrival of incident recorder Mark Styles who tells him, "None of your sarcasm."
[Kevin] "About what?"
[Mark] "The box!"
[Kevin] "Is your box infected then?"

Michael Slattery excitedly rushes over to greet Kevin, "I like your voice on the tannoy." Hilary Rouse stops to warn Kevin, "You wanna watch what you say to him, he writes it down all wrong."
[Jeff] "Jason Crump goes into the sex shop that Hilary works in."
[Kevin] "Really?"
[Hilary] "He don't come in any more cos he's moved to Sweden."
[Jeff] "He used to go into the sex shop that Hilary works in."
[Hilary] "It's not a sex shop. It's a top of the range high quality store."
[Jeff] "Shop you mean."
[Hilary] "Store."
[Jeff] "Jason Crump used to go into the mystery top quality store Hilary works in."

Hilary points to *Bouquet of Shale* (on sale for the first time), "If you put in that bit about the 'fat bastards', something I wouldn't say, there'll be trouble plus you put in about Freddie Lindgren – he's nice to speak to – about me looking at Freddie Lindgren's backside when it was Lyn [Francis] who admires that!"
[Jeff] "Don't you admire it then?"
[Hilary] "I admire Leicester. I've been to all the meetings except when I wasn't there yesterday. Alright, they don't win all the time but that's not why you go."
[Jeff] "It's the atmosphere."
[Hilary] "It's wonderful. They've all done really well with it. There's banking. I've got all the gear – except the jacket – because they haven't come in yet!"

Lyn Francis refuses to gainsay her admiration of Freddie Lindgren's nether regions but does express scepticism about Wolverhampton's presenter Ian 'Porky' Jones. She couldn't even sympathise when he collapsed on the centre green, "Never mind his knee, it's a shame his gob doesn't give out!" Hilary interjects, "I don't like the way he picks on the visiting fans all the time!" Kevin Long can't comprehend the geometry and load ratios of Porky's pirouettes, "I reckon that's a gravity defying feat to pivot on a small axis like that!" Michael Slattery scampers back over, "You're doing Ipswich again." Kevin pauses – for a fraction of a second, "Is that the word on the street?" While Michael listens intently he absent-mindedly wipes his fingers on the padded shoulder of Kevin's pristine cream jacket, "Don't do that! I'm not a hand towel you know."
[Michael] "I'm not."
[Kevin] "You are."

[[2] It's lucky that the tour didn't go ahead since Workington versus Ipswich also gets called off at 9am on the day of the meeting.

[Michael] "Have you got your microphone?"
[Kevin] "Yes, it's in my pocket. I'm not just pleased to see you."
The double entendre mystifies Michael, "I like you!"
[Kevin] "Johnny, do you want any help over there?"
[Michael] "I like you."
[Kevin] "I like you too [pause] really."
[Michael] "I like your voice over the tannoy."

After the first couple of months of a speedway season with new silencers, it's a genuine aural delight to hear some proper old school engine noise when the bikes start to rev and warm up 25 minutes before the meeting starts. One person's atmosphere is another person's noise nuisance. Though nowadays the number of city centre speedway tracks is a minority, the lengthy and arduous permission process experienced by Birmingham Brummies saw ongoing stringent noise control as a key condition of their successful application. Whether the fans want it or like it, FIM regulations mandate the introduction of said silencers in both the Elite League and Premier League this year (2011) while the National League have the dispensation of an additional year's grace to allow the less wealthy speedway riders additional time to save (prior to the across the board implementation of this regulation in 2012).

After only a few weeks, based at a track where deep shale is at a premium, Jon Cook embraces these changes with enthusiasm. "I was particularly pleased with the standard of racing and the amount of overtaking both inside and out, which flies in the face of some of the scare stories we have heard about the new silencers and I didn't feel the noise level was lowered to a level where one's enjoyment of the entertainment on offer was adversely affected." Ex-Poole Pirates press officer Gordie Day laments the road that took us to this situation. "We should have walked away from the FIM years ago. I'm sure Bob Dugard would agree if you asked 'what have the FIM ever done for speedway?' We should have formed our own association like Formula 1 or the MotoGP. They haven't got silencers and, if you chat to Boycie, he tells you that the sound of 30 bikes at the start of the MotoGP is unbelievable! You have to remember that the noise at speedway is part of the show! It sounds like they're doing 80 when really the average is below 50. Think about it? If you stand on the pavement and hear a car go by with an old exhaust, you think they're doing 70 – when it's only 35 or 40! Okay, we're different to the MotoGP because we're a city centre sport but the noise is still part of our show!"

Refreshed after his pre-meeting meal, Graham Rouse delights in the new-look Peterborough Panthers. "I'm still not sure if Nicki Pedersen is a good signing or not!? I like watching our three Poles – they give 110%! I think we'll finish mid-table but because we haven't ridden many fixtures, they soon pile-up and – if we get a few rain-offs – Nicki won't like that if he's doing well in the world championship. We'll have to see? It would have been an outrage – like it looked at one point – if there hadn't been any Coventry or Peterborough! I've been to every Leicester meeting so far, it's only 40 miles away, on a Saturday. It's got lovely long straights and banking. The track is getting better every week – it's settling down – and the crowds are good. They're certainly healthy enough for them to be making money anyway, which is good. Matt Ford won't like Emil Sayfutdinov going to Coventry because he likes to get everybody or thinks he should! On a false average, if at all possible! The rules shouldn't change because a small group thinks they should!"

'Eye of the Tiger' blares out over the speaker system ahead of the rider parade introductions by Kevin Long. Unlike the awful Derek Barclay visitor rider profiles that previously appeared in the Mildenhall programme a few years back, Phil Kirk's column provides both detail and insight. Kevin Long mixes levity with insight and updates in his introductions, ("a broken femur for Darren Mallett and Lewis Kerr is also out so we wish them both a speedy recovery"). We learn first about the visitors, in reverse order: Jake Knight ("a young man who is really sizzling in the reserve berth"),

Tom Stokes ("cousin of our own Lewis Blackbird"), James Cockle ("well-known figure round here as a former Tiger"), Oliver Rayson ("good season so far") and Scott Campos ("recent addition to the team"). "We'll have rider replacement at number 2 for the injured Lewis Kerr and there's one space still left in your programme – they had hoped for Kyle Hughes – but he's on a 28-day ban for officially withholding his services from Somerset – so we welcome Ashley Birks, instead!" People talk loudly nearby so I miss some home rider introductions but I do get to hear about Nick Laurence ("who joined the side last week and is keen to get amongst the points"), Jack Hargreaves ("hope for more from him") and James Brundle ("our captain who helped clinch that draw at King's Lynn with 14 on that occasion"). I also half hear talk about a mysterious but exotically named "Jojo Phelps", who's actually the less esoteric but useful Joe Jacobs.

James Brundle comfortably wins heat 1 after the challenge from Ashley Birks ends early. The second race includes the bonus of some alfresco kickboard riding-cum-testing. First Lewis Blackbird drifts wide to rattle the safety furniture mid-chase and also clatter the back-straight kickboard but somehow remains on. On the penultimate lap, third-placed Jake Knight enjoys a kickboard encounter that ends with an unscheduled dismount. Grumbles that the riders struggle with the track and their bikes continue. Dan Halsey goes from first to last and then shows impressive balance to remain seated after a bash into the back-straight kickboard. However, temporarily discombobulated, Dan immediately tests the fourth bend exit kickboard, falls, quickly remounts quickly but still gets a disqualification from SCB official Barbara Horley (with a trainee referee alongside). Heat 4 ends in dramatic premature fashion when, with Mildenhall poised for an easy 5-1, Jack Hargreaves crashes as he starts the last lap. Lewis Blackbird lays down and James Cockle manoeuvres past fallen riders and a loose rogue wildly careering speedway bike. Kevin Long offers praise, "Well done to the other three riders for having their wits about them!"

My refusal to buy a raffle ticket from seller Jacqui Cook prompts incredulity from John Louis, "He's got so much money, he should buy them all! He better or I won't buy his book again." Mike Moseley studies my book display and picks a £4.99 Track Directory, "This is about the right price for all your books – anymore is much too expensive!" Before heat 5, we learn the management of Mildenhall speedway are "delighted to welcome the students from West Suffolk Technical College." Who knows what they'll make of the meeting? After early race struggles with the track, the riders begin to look more comfortable with its condition. Like its predecessor, heat 5 is also awarded after third placed Daniel Halsey falls on the last lap.

Heat 6 sees an easy win for James Brundle to restore the Tigers lead, while Jack Hargreaves wins the next equally imperiously. After initial first lap pressure from Campos and Rayson, they race each other for second place for three laps. Aaron Baseby's confident wide second bend tucks him behind Tigers teammate Lewis Blackbird for their first 5-1 of the afternoon aided by an engine failure for Scott Campos. My ongoing memory loss sees me confidently greet Phil Hilton, "Hello Kev."
[Phil] "Kev? Kev who?"
[Jeff] "Long. Sorry it's the hair not the clothes."
[Kev] "I dunno if I should be insulted!"

Photography politics irks Phil much more than mistaken identity. "There's six photographers here today so I'm just hanging around the outside. Waller from Ipswich is here today – I've known him since we were 14 or 15 and we did cycle speedway together. I'd call him an acquaintance rather than a friend. On Monday before the Elite League [Pairs] at King's Lynn, he decided to come and send in his photos to the *Star*. You'd think they'd not put them in? Or that he wouldn't send them in for the 50 or 60 quid. I wouldn't do that at Ipswich. He should have his BSPA photographer's licence restricted to just Ipswich if he's gonna do that!"

Prior to the interval, Mildenhall pull further away with a 4-2 and a 5-1. A lady with two children closely studies my photograph book, "I used to come to speedway and now I've come back! I was forced to go as a child because my father was a speedway rider. You don't realise how much you enjoy it until you come back!"

[Jeff] "Who was your dad?"

[Lady] "I don't want to say."

[Jeff] "Okay."

[Lady] "Kevin Mullarkey – everyone has an opinion."

[Jeff] "I don't."

[Lady] "I went to Rye House a lot and all over."

Nick Laurence fires from the tapes in the rerun of heat 11 until brushed aggressively aside on the fourth bend by Ashley Birks (in black and white). He soon drops down the field to last place. King's Lynn's 2-7 restores vague respectability to the scoreline (38-31). Though arithmetically the Stars can fight back, the Tigers riders hit a groove they double underline with three successive 5-1s. Lewis Blackbird wins heat 12 with some élan and celebrates ecstatically afterwards. Heat 13 sponsored by "MSSC in Memory of Jean Mascall" is watched by her husband from the pits. It's a lusty race that sees James Hargreaves work his way from fourth to second with a lap to go. Ashley Birks refuses to back off until his engine fails. Young Stars team manager Dale Allitt sends out Oliver Rayson in the black and white helmet colour but he finds no way past Lewis Blackbird whose paid win gives him an impressive six ride paid 16 points from the reserve berth.

While we wait to learn the riders nominated for Heat 15, Kevin ties up the loose end of the raffle draw, ("prizes are £50, a bottle of wine and third prize is some biscuits – shortbread, we believe!"). Earlier Kevin outlined one of the dark arts of successful interviews. "A long time ago in this game, I realised that you have to ask open questions – otherwise you get closed answers. Then you get what I call a Ben Howe interview [pause], he gives yes/no answers to open questions!" While we continue to wait for the line-up, Kevin showcases his craft with a brief Aaron Baseby interview, "Do you feel any added pressure when it's a derby?" Never the most forthcoming of speedway riders over the stadium speaker system, Aaron leaves us in no doubt about the significance and status of this meeting, "Er, yeh!"

The drawn final race features a win for James Cockle and a first bend fall for Ashley Birks that earns post-race congratulation from referee Barbara Horley. The best time this afternoon is around 2.50 seconds outside the National League track record held by Adam Roynon (49.8 seconds) and the all-time track record held by Kenneth Bjerre (49.7 seconds). It looks a big crowd to me, especially compared to recent previous seasons. Johnny Barber disagrees, "I reckon it's the worst of the season, probably because it wasn't sunny." A 56-37 win sees the Tigers leapfrog King's Lynn into third in the National League table. Affable speedway fan Graham Rouse takes quiet satisfaction from an enjoyable afternoon of speedway. "Lewis Blackbird is 24. He rode, then he stopped and now he's back so he needs to get a move on. Leicester lost by 20 at Glasgow. Darkin got two! Last night they were trying to teach him English getting him to say 'ay up duck!' It'll be useful when he goes to Yorkshire. Mike Bennett was there last night for one night only. He kept trying to train the other bloke. He could do with some help. Last week he gave the heat 2 result using the riders from heat 3!"

**15th May Mildenhall v. King's Lynn (National League) 56-37**

# CHAPTER 2

# Ipswich

## "In fairness he was lucky to get that – I'd have given him minus one!"

19th May

Bathed in sunshine

The close-season distraction of horse trading about Coventry and Peterborough's membership of the Elite League overshadowed more significant changes within the Premier League. The 2011 season sees the arrival of an innovative but poorly understood later season PL fixture list plus changes to the speedway clubs actually competing. To relief in some quarters, Stoke speedway now compete in the National League while to much less relief Mike Bowden's Plymouth Devils move in the opposite direction. It's a move that apparently fulfils Mike's long-held ambition to progress the club far beyond its humble origins as a track in a rugby field. Ipswich Witches also went Premier League but were, possibly, quickly jinxed when tipped by pundits to be likely champions. Though Ipswich started like the proverbial express train, their smooth progress to the sunlit uplands of the upper-reaches of the Premier League has recently stuttered into the metaphorical railway sidings.

Said same experts also forecast a season-long duel between Plymouth and Newport for the booby prize of the PL wooden spoon. Even the most relentlessly optimistic Devils fan wouldn't expect an away win at Foxhall Heath, particularly without their talismanic number 1 Benjamin Barker (elsewhere at Sheffield in the British semi-final). Promoter Mike Bowden initially books Jason Doyle but his latest shoulder injury sees fellow Australian Cameron Woodward ride for Barker instead. Cameron had an early career PL apprenticeship with Edinburgh. Both teams also operate rider replacement on an almost tit-for-tat basis because of injuries to the Witches number 4 Morten Risager and Devils number 2 Lee Smart.

En route to the sandy sun-dappled Foxhall Heath grassland parking, I'm recklessly undertaken on the A12 by an old school minibus-cum-van with a "Devil's On Tour" sign prominent in the back window. From alongside said vehicle, the driver looks suspiciously like David Hawking. He films every Devils home meeting for posterity and, possibly, bumper DVD sales. Oblivious to his recent dangerous driving manoeuvre (and rather than glance right), the driver studiously concentrates on the road ahead as if passing a police car. Just ahead is a high-sided van with "No pies are left in this van overnight" on its back door.

With three hours before the scheduled start time, Chris Louis attempts to erect some signs above the entrance turnstiles without a stepladder. Clearly race-day stadium signage is part of the Louis' family promotional brief. In position above the firmly closed royal blue doors of the entrance turnstiles are large boards: one on the innumerable joys of the Speedworth Motorsports stockcar and banger car events and another full colour sign features an (unidentified) action rider photo with the inspiring strapline "Ipswich Sackers Witches Speedway no brakes, no gears no fears". There are also website URLs for both Sackers Recycling (visit their website to learn more about their "total waste management services") and the Ipswich speedway home page. The Witches URL – like speedway clubs everywhere nowadays – ends in the unusual but always eye-catching .co suffix. The BSPA are an early innovative adopter of said suffix (though this could be the equivalent of plumping for Betamax since I've yet to see any other company use it).

Though Go Speed International manfully battle to protect but rarely promote the (digital) speedway rights they own, nevertheless the technology revolution sweeps like wildfire through the speedway world. So much so that this season a slew of riders, journalists, and other notable figures within the sport join Twitter to share their thoughts in 140-character bite-sized observations. These tweets range from witty to banal via indecent, obscene and self-aggrandising. Greg Hancock adopted the medium early so has more tweets and followers than any other contemporary speedway rider. He's a serial, almost obsessive tweeter. The speedway Twitterverse might soon take a revolutionary leap forward since Nigel Pearson taught Kelvin Tatum how to tweet on a recent overseas trip. Though the end result of this tutorial-cum-masterclass is yet to bear fruit, given how many inappropriate-isms and neologisms that litter Kelvin's live commentary work, he's got no real reason to feel self-conscious about sharing any spelling mistakes or platitudes with the wider world. While Chris Louis continues to size up the stadium entrance signage – and though he doesn't strike me as the type of man to court the pitiless glare of speedway celebrity – it's an ideal moment to check out Chris's enthusiasm for the Twitterverse. "Nigel's been teaching Kelvin how to tweet, are you on Twitter?"
[Chris] "I am – for the club. Well, I don't but someone else does."
[Jeff] "Are you going to tweet for yourself?"
[Chris] "I don't twit. I can't be bothered with it, I'm afraid."
Over in the Ipswich speedway office, promoter John Louis meets and greets a steady stream of arrivals. With a box-fresh copy of *Bouquet of Shale* in hand, he asks, "How many do you sell at each track?"
[Jeff] "It varies, I'd like to think five to ten."
[Chris] "Is that all? Do you sell some by post too?"
Clearly it's too late to add the caveat that – recently – these sales numbers would represent a good day.

In the kitchen area adjacent to the Ipswich speedway office, there's another deliciously moist fruit cake freshly baked by Chris Ellis. Between race-day queries on the club phone, Chris gazes out of the window towards the pits and track where her husband Bob works hard in the sunshine. Matter of factly Chris advises Chris Louis, "There's a lot of people out there without passes". Chris shrugs, "Let it go this week but if they're not doing anything, they're no use." Chris also wears a metaphorical Mildenhall Tigers promoter's cap, "If we put on the American touring team – the good one with the young ones, not the old – do you think we could get a crowd over from the airbase?" Though time-rich bored American service personnel with disposable income lurk behind the perimeter safety fences and barbed wire, they remain the dog that doesn't bark with regard to regular trips to West Row stadium to watch the speedway.

Smartly dressed Witches presenter Kevin Long wears his trademark tailored white jacket, immaculately ironed shirt and tie, while his sharp creased trousers match his quick wit. The kitchen window continues to get Playschool levels of attention. Kevin spots tonight's SCB official Dave Robinson in the pits, "Oh God, how come we've got him again after what happened at Rye House? He got two decisions totally wrong and cost us a point. I went up to the box afterwards and told him. First of all he stopped heat 13 when Frampton threw himself off and broke his collarbone. Frampton fell behind our two who would have changed position over the next lap and a half. He said, 'They had the bend and the back straight to sort themselves out' before he put the red lights on. But how could they know he'd fallen behind them? He then said, 'They might not have changed positions.' I asked him how long he'd been watching speedway because I can count the number of times on one hand when riders on a tactical don't swop position on the last lap! Funny enough, we had one here last week. Then, even worse, he declared the meeting after 13 heats when we could have still got a point, if we'd done okay in the last two heats. He then said, 'I don't have to explain myself to another Ipswich official because I've explained myself to Pete Simmons already.' I said 'I'm here in a personal capacity and paid to get in!' He asked, 'What are you doing in the box then?' I told him I'd knocked and he said I could come in!" Chris Ellis takes a dim view of the financial cost to the club of the forced cancellation of supporters club tour planned for this weekend. "We've lost £500 on the cancelled tour. Chris [Louis] is going is ask the BSPA for the £500 we lost. The Premier Inn in Scunthorpe were fantastic – they gave us our £1,455 back – then they do special offers for speedway fans there! Whereas the Swallow Hotel at Carlisle kept back £500 because we cancelled between eight and 14 days beforehand." In addition to the thrill of the Devils visit to Foxhall Heath, tonight features the additional attraction of British International ice speedway rider Kevin Rowand's attempt to break Hans Andersen's track record of 57.7 seconds. Kevin Long has this news on his handwritten presentational aide memoire, "Kevin Rowand is going to go round in the interval and try to set a new track record – because he can! As long as he don't go anywhere near the airfence, we'll be okay!"

Supporters' club chairman and programme contributor Peter Thorpe clutches his notebook and Papermate pen embossed with the fantastic words, "From the Chlamydia team". This should be a story in itself but, instead, our conversation centres on the camera slung around Peter's neck. "I don't think Seemond is going to enjoy how grippy the starts are here at Ipswich! He might have done well in the National League but he's just not going to be quick enough tonight. I don't expect it will be a big crowd but I've brought my camera so I can get a photo of Seemond and Mark Simmonds." Aside from his admiration of returnee riders, Peter contributes weekly to the programme. "I'm doing a column and I was going to do one on each rider and I thought that would cover seven weeks but I've so much material already some riders will be in for seven weeks!"

Affable Harry Bushnell often supervises the spares van at Mildenhall (sometimes wearing speedway gloves) and is amazed how some riders can even afford to ride. "Do you see John and Craig Branney have retired? They were Premier League level riders and yet they can't afford to continue! I don't know how riders who aren't as good as them can continue riding when you think that steel shoes was £67 in 2010 and now they're £84. A tyre is just over £43 and what with methanol at £8 a gallon you need a couple of those for a meeting!" Though Sky television cameras at almost any live Elite League meeting decimates attendance levels, it's often justified as a potential shop window to attract additional race-day support and interest for speedway clubs. As if to illustrate the point, Chris and Von Laughlin stop by my table on their first ever trip to a speedway meeting! "Excuse me, where we should we stand to really enjoy it?" They're fortunate to choose Foxhall Heath since this is a circuit where you can watch with many good vantage points within the stadium. For atmosphere, proximity and drama I suggest they watch from the grandstand that overlooks the start line. "We watch it on Sky and really enjoy it! So Von suggested that we come along and see what it's all about live. We wouldn't have come but for Sky." Chris has further questions. "Will we be able to move about during the meeting if we want? How do we use the programme? Do they have set rules?" Chris and Von intend to take full advantage of the casual freemasonry of speedway fans surrounding them to ask anyone on hand any questions they have.

The Noah's Ark style rider parade that sees rival riders circuit the stockcar track in numbered pairs prior to their introduction by compere Kevin Long is an Ipswich tradition. We're told about Seemond Stephens, "It's says in the programme 'Another Cornish legend' I don't know if that's a good or a bad thing?" Kevin also bigs up the interval track record attempt by Kevin Rowand but emphasises the safety preparations, "we've got a thousand puncture repair kits ready!" In the programme booth (with Chris Ellis), between sales Jason clues me in about his speedway travels. "My two favourite tracks are Scunthorpe and Somerset. It's weird Scunthorpe works – because of the adverse camber – but it does! I asked Carl Wilkinson about it and he said that you don't ride over the camber but round it and then back up onto the track."

Kevin Doolan's easy Heat 1 win ahead of Devils guest Cameron Woodward doesn't bode well for a competitive fixture. Afterwards Kevin Long snatches a few brief words with the likeable and articulate Australian, "He made the mistake – I went on through and built up a good bit of speed there." While Doolan returns to lurk meaningfully by his bay in the pits, Kevin Long flags up the lesson of the interview for any slow-minded fans, "An interesting insight into what goes through a rider's mind during the race there!"

Heat 2 starts promising for James Cockle and Jasper Kristiansen when they both gate. Sadly for the very few Devils fans here, that's almost as far as it goes after James Cockle gets it all wrong and crashes to lose the lead. Continuing in schoolmasterly fashion, Kevin Long draw further lessons for the crowd, "As we just discussed with Kevin Doolan, a lapse in concentration can lead to a massive unforced error!" Though the airfence remains intact, James Cockle may be less so. "We need to delay the start of heat 3 because the paramedic has gone into the pits to check that James Cockle is okay." It would be hard for Ipswich team manager Pete Simmons to give a clearer or more savage indication of his damning assessment of the likely level of threat/challenge presented by the Plymouth "Stitches Tailoring Protyre" Devils tonight than his decision to use Witches number 8 Lewis Blackbird as rider replacement ride (for Morten Risager) in heat 3. It's a race only really notable for the sight of Seemond Stephens massively off the pace – apparently in an intractable but passionate struggle with his bike, psychology and the track. Seemond's so far behind that he can't recover third even after Lewis Blackbird makes a comprehensive hash of his third lap. Heat 4 sees Jason Bunyan follow-up his bike breakdown on the warm-up lap with a first lap engine failure. James Cockle is soon so far behind Messrs Schramm and Mills that he prefers to ride upright – rather than race – for the last lap and a half.

At the trackshop, the commemorative "£2 real postage stamp" to celebrate the Ipswich Witches 60th Anniversary 1951-2011 catches the eye. Skilfully combining his lifetime in speedway promotion and merchandising with stamp dealing, Colin Barber produced this unique memento for similarly minded speedway philatelists (with access to the Ipswich trackshop). Colin's son Johnny is at work in the shop with mum, Molly. "You've got me in trouble with my brother thanks to your book. I didn't look at the caption and said to his two daughters, 'Look there's a picture of daddy in this book!' When Nick looked at it he said, 'That's Rob Peasley!'"

Devils team manager Gary Spiller tries out the black and while helmet colour on Cameron Woodward only to see him finish second behind Lasse Bjerre in heat 5. In contrast to Plymouth (who have their captain Mark Simmonds take Lee Smart's rider replacement ride), Ipswich again send out Lewis Blackbird for Morten Risager's in the full knowledge that this almost inevitably gifts Cameron double points. While we await more race action (rather than a contest), Kevin Long excitedly name checks the club's main sponsor (Sackers) and reminds fans they offer "full service waste management services". Quite what exactly these services are – politely, I imagine – isn't specified? According to Johnny Barber, who admires the crowd size, "There must be 1,200 here which, if you'd been here for some meetings last season, would count as a bumper crowd!"

Before heat 6, Jason Bunyan fails to line up and is disqualified under the time allowance since he's back in the pits with mechanical difficulties. It's news that creates more envy than sympathy in the fan stood next to me, "Bunyan has that New Zealand Rose as his mechanic – she's the most attractive mechanic in speedway!" James Cockle replaces Jason and in the rerun and – for the second time in three outings – fires from the tapes only to zoom straight into the airfence with a throttle jam. This incident prompts further sang-froid, "It's carnage here tonight! That's the second time Cockle's crashed! Never mind that Bunyan's bike broke down in the warm-up, that he got as far as bend one first time out and now he ain't out here either!" There's considerable further delay before the rerun of the rerun. It's a delay that mystifies Kevin Long ("not sure what is happening?") and exasperates my neighbour, "What are they doing here tonight?" Kevin reminds the crowd about *Bouquet of Shale* which, apparently, comes "from a slightly different stance!" He continues, "Jeff and I tried to work out what a series of six books is called, if three is a trilogy? We couldn't work out what that was so if there's someone more intelligent in the crowd than us, please let us know." When heat 6 finally runs to completion (with only three riders after Cockle's second disqualification of the evening), Jesper Kristiansen surprisingly relegates Kevin Doolan to last (third) in a race won by Taylor Poole. Heat 7 is another race only really notable for Seemond Stephens trailing off at the rear, albeit even more spectacularly than last time out. Colin Barber draws deeply but phlegmatically on his cigarette, "Seemond looked poor in his first race and that one. He's always been dreadful away! He was when he was at Exeter and he is now!"

Before heat 8, after two crashes James Cockle withdraws from the meeting. When the race gets underway, Jason Bunyan finally completes four laps but – from the second bend of the second lap – isn't really competitive. Colin Barber shrugs, "Jason two-wheeling into the corners, eh?" Away from worlds of philately and speedway, Colin sometimes watches the Dogs. "I heard Sittingbourne had permission for four years. I was watching the greyhounds last week and Roger Cearns was interviewed about their struggle there and he said he'd taken a pay cut but also that speedway would be there!"
[Jeff] "Don't they have an early curfew?"
[Colin] "It was said to be 8.30 but I think that was wrong. It's probably 9.30 so they could start at 7. People say they could get 4,000 in. That's silly! It's bigger than here and they can get 10,000 in here. It used to be 20,000! They might start in the big league but I think that might be a mistake – they should try it out in this [Premier League] first. They say Terry Russell looked. It's a magnificent stadium! My friend Wally Mawdsley looked years ago." Elsewhere on Foxhall Heath, Kevin Long grabs Chris Louis about their abandoned Northern Tour. "It's a crying shame that the tour had to be cancelled."
[Chris] "Yeh, a big disappointment for the fans, the club and Chris Ellis! She always organises a fantastic trip."
[Kevin] "Don't the authorities take the fans' club tours into account when scheduling meetings?"
[Chris] "It doesn't appear so!"
[Kevin] "I think we better stop before it gets too political!"

Heat 9 effectively ends as a contest after Lasse Bjerre casually leans past Jasper Kristiansen on the second bend. Jason Bunyan records his first point of the night ahead of Witches number 8 Lewis Blackbird (out for his third experience-building ride tonight). Kevin Long marvels, "Incredibly that's the first shared heat of the night!" Our collective incredulity levels ramp up further when we learn Lasse's winning time of 58.5 seconds establishes a new Premier League track record that's only 0.8 seconds outside Hans Andersen's track record on this reconfigured version of the track. Kevin Doolan finishes third for the second time in three races in heat 10 ahead of Seemond Stephens who – once again – brings up the rear. Johnny Barber's seen enough, "Seemond Stephens looks – can we say, putting it politely – ready for retirement!" Kevin Long snatches a few words with the often plain-spoken Lasse Bjerre, "Last week you took a while to warm up and then you were flying. This week you're flying from the start and you're winning your races by the length of the straight!" Lasse modestly bats off the praise, "Yes, it's the first time they've done such a good track, I like!" Kevin's quick to defend this oblique criticism, "Hold on, Bob – our trackman – always says he has 14 track curators each week – seven on each side!" After a bit more chitchat, the interview concludes. Announcer Mike Western interjects, "Well done, no swearing – excellent!"

Between heats 10 and 11, referee Dave Robinson grants additional time to Seemond Stephens only to then disqualify him under the two-minute allowance when he fails to take Lee Smart's rider replacement ride. When the race gets underway, Cameron Woodward is Plymouth's first race winner here tonight. An incredibly loud pre-recorded message serenades in our interval break, "We are BBC Radio Suffolk – we love speedway – especially local speedway!" It continues loud and breathlessly and we learn that someone called Stephen Foster "is the voice of Suffolk speedway".

After a season when the pros and cons of the new silencers are cause for debate, I'm afraid to report that the muted new version silencers don't quite meet with my approval (just yet). That said, Kevin Rowand ruins the final few heats before the interval. He stands posily revving his ice speedway bike on the stockcar track adjacent to bend one. Speedway is always better without the irritation of this additional accompaniment. After much fanfare, Rowand comes out on his customised ice speedway bike. Like the standard speedway bike, it's also 500cc and runs on methanol but, significantly, unlike them has two gears. As a consequence of said gears, Rowand makes a slow start before he gains enough momentum to take the corners at an unfeasible angle. With four laps complete, Kevin Long expresses satisfaction that the safety furniture remains damage free, "I'm pleased to see the airfence survived intact!" Mike Weston reveals, "58.6 seconds. Well, it was a quick one but it didn't break any track records!" Back in the pits, Kevin Long is anxious about impressionable teenagers, "Did that scare you?" Taylor Poole admits, "It's crazy!" "Bet you don't get much ice in Australia, do you?"

During the post-race interval interview with a laid-back sounding Kevin Rowand, we learn that he's from Plymouth and, thereby, massively boosts their fan numbers at a stroke. Between customers, Johnny Barber studies the speedway updates website news from the British Final qualifier at Sheffield. "Oh no, Chris van Straaten's plan is coming unstuck tonight! Chris Harris only had four points from three rides and the Cardiff wildcard is Tai's!"[1] Adie Fowler is underwhelmed with tonight's action, "It's one of those meetings that you just want to be over so you can say you went!" Just prior to heat 12, Johnny Barber races from the trackshop towards the first bend to catch the race action. "Come on Jeff, it's the big one! Lasse Bjerre versus Seemond Stephens – you could see your first lapping! Seemond was excluded from the last one because he was finishing the race before." Seemond avoids the ignominy. Though comprehensibly last, after three laps, Johnny concedes, "I'd almost say Seemond was competitive in this one!"

Heat 13 sees Jason Bunyan complete another race (his second tonight!) but, again, he fails to trouble the scorer. Chris Ellis takes a philosophical view, "It's good but they haven't got much of a team. It's a shame!" Though Devils team manager Gary Spiller sends out captain Mark Simmonds for a tactical ride, he's unable to beat Jerran Hart, though, with Lewis Blackbird out yet again as rider replacement, the Devils get their second 3-5.

From behind a centre green lectern that makes Kevin Long appear part man of the cloth part teacher, anticipation levels reach fever pitch ahead of the big raffle draw. "Lewis Blackbird is going to do the raffle draw. You just don't come along for the ride! Don't say shades to me, its pink or nothing at all! It's not cerise!" In the last race of the night, Lasse Bjerre finishes third to see his hope of a five ride maximum evaporate.

[1] Chris Harris would buck up in his last two rides to thereby qualify for the British Final at Wolverhampton.

With 13 Ipswich race winners, it's a surprise the score is only 59-35. Having intrusively accompanied the last four races of the meeting with the irritating drone of his ice speedway bike, Kevin Rowand makes another attempt at the track record. "Right boys, compose yourselves – we've got dangerous work going on again!" This attempt does look quicker, "I can tell you that Hans Andersen is quaking in his boots!" Mike Weston delivers the bombshell news of a new track record of 57.6 seconds (0.1 seconds faster than Hans Andersen). Kevin remains slightly sniffy, "I bet he can't do donuts on it though!"

Well-travelled Ipswich speedway fan Mark Shrimpton tells me, "It's a shame Ben Barker didn't ride." During the close season, Mark went to Argentina to watch speedway! "It was an absolutely fabulous trip. I wouldn't say I'd rush back there but we managed to get three and a half meetings on three tracks. The first one was rained off. From another, we got an overnight coach back. We had to book ourselves on the 02.30 coach because the 03.15 was full, so we had to leave a few races early. It wasn't due to finish until a quarter past one!" Plymouth effectively tracked a three-man team of Cameron Woodward, Mark Simmonds and Jesper Kristiansen. Due to mechanical difficulties, Jason Bunyan scores paid two while Seemond Stephens gets nothing. Johnny Barber shrugs, "In fairness he was lucky to get that – I'd have given him minus one!"

19th May Ipswich v. Plymouth (Premier League) 59-35

## CHAPTER 3

# Rye House

## "They've not got their ginger-haired Australian"

21st May

Home straight

The end of the world forecast is for 6pm. Everyone copes in their own way. Going towards Brighton is a red van sprayed with the words 'Pussy Wagon'. Judged by the look, décor and age of the vehicle rather than the signage, the driver is an optimist. In a hot, dusty Hoddesdon car park, the Rye House fans already there wait patiently with clipboards, fold-up chairs, flasks and sandwiches in preparation for Rye House 'Silver Ski' Rockets versus Somerset 'Dickies' Rebels rather than the forecast Rapture. With bright sunshine, light breeze and a cloudless sky – unless the world ends, the meeting will go ahead. With the turnstiles yet to open, Gaz can safely desert his race-day refreshment kiosk duties, "Oh God! He's here again! Watch it – he'll write down everything you say. Am I innit?"
[Jeff] "Yes."
[Gaz] "Saying what?"
[Jeff] "What you said."
[Gaz] "What, that you're a tosser?"
A tall tanned speedway rider with an Australian accent takes a close look at my display table, "Are they speedway books?"
[Jeff] "Yeh."
[speedway rider] "You write them?"
[Jeff] "Yeh."
[speedway rider] "Ah, cool!"

Andy Griggs's trackshop helper sees retro echoes in the young Australian's fashion sense, "Nelson is gonna get the shades to really look like Andrew Silver! He's already got the kevlars tucked in his boots."

Sam Masters and trackshop franchisee Andy Griggs chat animatedly about future merchandising plans, while the tall tanned young Tyson Nelson – Rye House Rockets reserve rider on an assessed

5 point average – waits. From the programme racecard, it's clear he's dropped his first name and prefers to be called Nelson. Promoter Len Silver's expresses surprise about other matters in his column (Len's Lines), "Nelson is struggling to find an engine that suits his style of riding". The recent away meeting at Newcastle saw both Luke Bowen and Steve Boxall return to the team. Apparently, Steve's return in place of Jason Frampton prompted ill-informed speculation on the speedway forums. Len sets the record straight, "A few people need to learn the FACTS about Steve's re-entry into the team. At the beginning of the season Steve was listed and registered for the Rockets. I had been totally satisfied that he was very fit and determined to do well and with good machinery to ride. Then he suffered that arm injury and other complications, which stopped him from riding. So I was lucky enough to sign Jason who was out of work at the time (though not for long as Poole needed him just a day or two later). The deal with Poole, Jason and the BSPA was that Jason would be a Rocket for as long as Steve was out of action, but that switch had to be reversed as soon as that situation changed. So for those few who thought I could keep both riders, it can be seen that this was not possible."

Discarded cotton wool flutters in the air and drifts into small piles against the concrete wall adjacent to the sand-covered ex-dog track that separates us from the speedway track. The raffle lady Shelly, explains, "It's the willows, they're all on this side of the river. We've had it for about three weeks now, it's nearly over. It's like cotton wool or snow, dry snow". The turnstiles open at 5pm, "You get a rush of the normals – all the people who travel a distance who like to come in straight away and then the real rush starts at 6.15pm until about quarter to seven." After the initial influx with the turnstiles temporarily becalmed, I go over to take photos.
[Michelle] "Why do you want to take our photos?"
[Irene] "Is it for the *Speedway Star*?"
[Michelle] "Is it because we're the friendliest?"
[Jeff] "Yes."
[Michelle] "We are. It's better to be that way. You get the best out of people, there's no point being otherwise."
[Jeff] "There's a nice breeze."
[Michelle] "There is. Usually it's freezing cold but, today, it's alright."
[Irene] "It's blowing in all the bits."
[Michelle] "You're lucky you've come now while it's earlier. We complain about it, whatever it is! Either it's too hot, too cold, or too windy. If you come back at quarter to seven, when it's really busy, you can hear all about it."

Passing the raffle stall, I overhear, "He's got the physique of a body builder!" As if on cue, Craig 'Mr Potassium' Saul arrives clutching two bananas, "I've got them specially for you." Craig's son Dan denies this claim, "You bring them every week." Craig puts his fruit to use. "When I arrived at Peterborough last week, I had one banana with me and I'd left the other in the car. Peter Oakes rushed up and said, 'Linus Sundström needs a banana for his pre-meeting meal.'"
[Jeff] "Why didn't he have one with him if it's his pre-meeting meal?"
[Craig] "I dunno! A lot of them have something greasy or drink a can of Red Bull. Later at heat 10, I said to him, 'There's no such thing as a free banana! You have to do an interview.' He was fine about it."

Only out a few days, Craig's already read quite a few chapters of my new book for interest and research. "I must get a second copy. I keep the other at home for best and travel with the other or lend it to people – so I know it'll come back dog-eared. Now that I'm at Lakeside, I get to see Scary Sheila la Sage every week. I didn't know about her until I read your books. When they said 'Scary' over the speakers until then I thought it was a name that shouldn't be mentioned. I also see Wendy – Miss Fina Invader – at Peterborough."

[Jeff] "Ah! Wendy Jedrzejakski."
Ever professional and alive to the nuances of pronunciation, Craig says, "Is that how you say it?"
[Jeff] "It's how I say it."
[Craig] "They must have shut in 1997 or 1998."
[Jeff] "I'm too polite to mention that, you can!"

Somerset programme editor, press officer and webmaster Ian Belcher temporarily leaves the pits. "It's our twelfth official meeting tonight and we've only fielded our top seven twice! That was during our first meeting of the season – even then Jimmy Holder crashed on his first ride on the third bend in heat 2 – and on Sunday, at home. We're lucky with Travis – he's got a high average – so we can have a guest because he's number 1. Dakota's will go up shortly. Rolling averages will work if they keep the system going."
[Jeff] "They'll abandon it in the winter."
[Ian] "Or sooner."
[Jeff] "Are you going in to win tonight?"
[Ian] "Depends which Rye House team turns up! The one that beat Ipswich or the one that lost to Newport at home! We'll do alright if Ty Proctor turns up! It's quarter to six and he's not here yet!"
[Jeff] "Are you going to all the meetings again this year?"
[Ian} "I'm not – I'm going away – missing eight matches."
[Jeff] "Eight!"
[Ian] "Yeh, it's best to take the other half away. We waited and waited and then booked a holiday in France. We both like the area where we're going – it's the Loire Valley – and can take the dog. Typically they don't publish the fixture lists and lots of meetings are on then – including the Scottish tour! I'm going to do a programme before I go and update the website while I'm away."

Like many Premier League fans, Ian struggles with the concept of the new 2011 structure. "The second half of the season is all wrong! You could end up with a situation where Glasgow gets Leicester, Somerset, Plymouth and Newport so they lose all the local derbies and get all the travel costs. Whereas, if at the start of the season, when all the teams are supposedly equal, they could split it into North and South, so you have all the local derbies at the start of the season and then again in the second half. They would also be able to publish the fixture list, which we don't know yet."
[Jeff] "Isn't it first, third, fifth?"
[Ian] "Nah, it's not odds and evens, it's more random than that! If you look at our website – Andy Povey does it – it has a breakdown. I know it goes second, fourth, fifth – that's the first bit anyway."
[Jeff] "Do you reckon you'll beat Leicester?"
[Ian] "When you come next week we'll have a bit of a Select side because Dakota North and Sam Masters will be away at the Under 21s and, hopefully, Travis will be back, we'll have a guest and Christian Hefenbrock or, maybe, not!"

With some gusto a familiar faced man bounds up. After friendly hellos he says, "You don't know who I am, do you? You usually see me at Sittingbourne. I must apologise for last year because, when I read the book, you had put in nice things about me. Sorry I doubted that! In addition to doing the reffing there, I sponsor the Amateur and Southern rider meetings. I'm the MTD Group – not many people know that or notice my number plate – JS5 MTD which gives it away!"
[Jeff] "No one is that observant."
[John Strong] "I'd rather sponsor the Amateur meetings where what I do makes a difference rather than do something elsewhere, which would be a drop in the bucket. I'm the incident recorder here, which I enjoy too!"

Craig Saul's dulcet tones burst loudly over the stadium speaker system, "A very sad Sunday –

whoa – I'll put my teeth back in and start again, a very sunny and happy Saturday!" Craig bigs up the meeting ahead, name-checks all the facilities on offer inside Hoddesdon Stadium including the bar, acclaimed fish and chips as well as the trackshop ("all your Rye goodies from posters to pens and clothes"). Two smartly dressed women rush up with an urgent question, "Excuse me! It's our first time, can you bet on the races like the dogs?"
[Jeff] "No, you can't."
[Lady] "It would be good if you could and a good way of making some money!"
[Jeff] "You could bet among yourselves."

Reading speedway historian Arnie Gibbons is a man you can rely on for (questions and) answers. Is the predicted end of the world Rapture at 18.00 hours East Coast or West Coast time? "No, it's 6 o'clock local time. I was listening to Radio 2 on the way here and, when there was the handover, Alan Carr said, 'I haven't got anything prepared for the show!'" Arnie's History of the Reading Racers is a critical and sales success, "When I closed the accounts on my book it had made a profit, which I've used – in part – to sponsor Phil Morris's testimonial in September and the Under-19's trip to Sweden!"

With his presentational work still to properly get underway, Craig Saul listens to Arnie's forthcoming concert plans, "I'm going to see John Cooper Clarke and John Otway at the Royal Festival Hall. The last time I saw John Otway was at the Lucas Aerospace Sports and Social Club in Hemel Hempstead!" Music at speedway meetings is variable. Arnie's international travel takes him to obscure places, "The worst evening's music at speedway ever was over at Grindsted (in Denmark). They played Smokey's greatest hits and, when they finished, they played it again!" I like to hear about embarrassing gigs. Craig confesses, "I went to see Can play at Bournemouth Winter Gardens. There were 25 people in the audience only and that was embarrassing – they came on 40 minutes late." Arnie recalls, "As I'm Reading born and bred, I went to the Reading Festival and saw Wayne County and the Electric Chairs. This was before he became Jayne. They sang two songs, '(You Make Me) Cream in my Jeans' and 'Fuck Off', I think, before a hail of cans drove them from the stage."
[Craig] "At the last night of the season at Peterborough in 1993, Rod Colhoun had been dropped from the team and he brought his guitar along and played it in the bar. He played a spoof of 'Knocking on Heaven's Door' called 'Knocking on Kevin's Door'."
[Arnie] "You don't want to be standing next to David Miller at the Royal Opera House urinals. Your mind turns to things you don't want to contemplate! The timings work very well of Glyndebourne and Eastbourne speedway together! It starts an hour or two before the speedway and finishes at almost the same time. So I can drop Paul off on the way there and pick him up on the way back."

Craig's duties call him away. Arnie's got no sympathy with people who doubt the legitimacy of Peter Adams's recent double use of tactical rides as rider replacement. "The rules over using tactical rides in a rider replacement are perfectly clear. As the brains trust on the British Speedway Forum points out, it's not ruled out, so it is legal! A better question is, as a guest is only allowed for a number 1 or a British entrant in the Under 21s, so how did Leicester get a guest for Richard Sweetman a few weeks ago when they should have had rider replacement?" Arnie's joy at the return of Leicester is unalloyed. "It's only 80 minutes to Rye House from my house and 2 hours 5 minutes to Leicester. Going back is all about reliving my glory days. I'm standing there chatting with people I stood with 28 years ago! When you go, watch out for people with Ray Wilson rosettes. Apart from the fact the racing has been a bit disappointing because the tracks been a bit hit and miss, I think the problem – and I'm not an expert – has been the extraordinarily dry weather, which has made the track like concrete with a shale top surface. If you water it now, it's concrete and it runs off. Most people have been tolerant and understanding. At least they avoided the Wimbledon problem of a first meeting of sludge. The other thing is the announcer. It will disappoint you to hear that he probably won't be there when you go for the Berwick meeting. He's been announcing race results in random order. Fourth places and a 4-1. Telling us who was fourth was one thing but he also had a tendency to announce riders not there and to forget programme changes. Promoter David Hemsley has the cycle speedway connection so they're getting in the cycle person. I went to the first away meeting and that was really moving! I went to the first practice meeting and that was extraordinarily uplifting! There were all those people I used to stand with or know. The last time I saw Keith Hallam was at your table in Peterborough in 2006. He goes with his dad who's in his 80s. We're the same age – 51. I appear to be two weeks older than Derek Barclay. And there's Elmo who was an impressionable 16-year old when Leicester packed up in 1983, now he's a parent and I have to think of my language!" Arnie isn't fussed with the furore over the new silencers, "I must admit I only notice it if I consciously listen! It must be an extension of being tone deaf."

Veteran speedway promoter and inveterate showman, Len Silver takes to the microphone. "Welcome to our good friends from Somerset and the people from Somerset – hello my dears, hope you enjoy the meeting." After various further greetings and comment, Len soon gets down to rider introductions. "They've not got their ginger-haired Australian [Travis McGowan] instead they've got a very clever guest from Wolverhampton – and the Wolverhampton track is almost identical to ours – Ty Proctor. Dakota North – their number 2 – has made a difference with a surge of form showing that he's a champion. Christian Hefenbrook – he's the German who's broken his ribs so they run rider replacement. Sam Masters – a rider who loves our track. Cory Gathercole – another Aussie: they do love them from Down Under! Alex Davies – I think they got that one just past the BSPA – and James Holder, we saw him here for Ipswich".

Len continues with his own team, "Number 1 as always, Jordan Frampton!" Arnie interjects, "It's not always – it was Jason Doyle!" Echoing his programme notes, Len welcomes back Luke Bowen, Ritchie Hawkins ("getting better week by week"), Chris Neath, ("Mr Rye House"), "Let's give a big welcome back to the team for Steve Boxall [big cheer] and let's also welcome our reserves: Ben Morley and Nelson." Arnie tuts, "I do hate this affectation – his name is Tyson Nelson!" After the formality of the toss for gate positions between captains James Holder and Chris Neath, Len traditionally throws the "lucky £2 coin" to the gaggle of youngsters who gather by the start line for this very purpose ("not much blood tonight – shame!") Arnie sighs, "I get the impression that the locals regard Steve Boxall for Jason Doyle as strengthening the team!"

After an amendment to our programmes ("there's one typo in heat 2 – Chris Neath isn't riding, it's Ben Morley"), Craig Saul gives a warm Rye House shout-out welcome, "to the Royal Mail and the Hertfordshire Constabulary who are guests in the hospitality suite tonight!" After a drawn heat 1, Craig completes his race line-up introduction for the next race with, "The sheer guts and explosive excitement – yes, yes, it's Nelson!" The race fails to get meaningfully underway because James Holder demolishes the tapes, "there's movement at the line there – there's disqualification for James Holder – a visitor here with Ipswich recently – for finding, nay breaking, the tapes there!" In the rerun, Rebels reserve Alex Davies completes a comfortable victory to admiration from Craig, "making a very promising debut there – Alex Davies with a time of 57.9 second for the 18-year old." Effortlessly knowledgeable, Craig tops and tails each race with facts, insight and information in an unobtrusive but useful fashion. Chris Neath takes Sam Masters right out for a close inspection of the third bend fence before he zooms away. "I tell you Chris looked fast in that one! He wasn't fast enough to beat his own track record of 55.1 seconds, but he was fast at 55.6 seconds".

Judged by the racecard, heat 4 pits Steve Boxall ("it's our first look at the rider in red. A re-debut, if you can call it that, after that meeting at Newcastle") against Cory Gathercole ("a third year Rebel after two seasons on the Isle of Wight"). The reality is different since Boxall and Nelson hammer home ahead of James Holder. "Winning on his debut again, I give you the Boxmeister

– Steve Boxall!" This is a win greeted ecstatically by a small knot of Steve Boxall's family and friends gathered on the steps outside the bar. Steve Boxall's mum asks the young boy next to her, "Do you like it then?" He does! Plus he's got some good advice of his own, "Always red or blue, always!" The advice blows freely, "I tell him: everyone else don't care about your teammates! They're professionals, they can look out for themselves." Gaz warns another unsuspecting fan, "Don't stand here, he'll be writing every blooming thing down. He is a flapper-jacker!" Heats 5 and 6 are drawn after reruns. Luke Bowen's falls and disqualification from heat 6 sees him take no further part in the meeting.

The night suddenly takes a possibly promising turn when Hester Adams asks, "Are you Jeff Scott? I came tonight because I saw you were to be here on your website. I came here specifically to see you! The website said you'd be here. Dunno how I got there? I brought my son Eddie – it's a break from his GCSEs – he's over there. I think he's quite enjoying it." Away from the track Hester is a councillor. She also freelances as a television researcher, hence her trip to Rye House speedway tonight. However, from the scant details I gather about the programme Hester researches, it's a reasonable assumption that someone at the television company reads the local newspaper reports. "Someone had an idea in the bath and thought speedway would be a great setting for a Channel 4 crime programme! It's nothing more than an idea at the moment. I can tell you no more than that, but I wondered if I could ask you some questions?" For the next six or so heats – between the race action and my vital programme work – Hester quizzes me and takes copious notes. Though speedway can shoot itself in the foot, Hester soon learns we keep our real/imaginary dirty washing within the strict confines of the immediate speedway family.

Similarly, no matter how many times I press Hester for further detail about the conceptualisation and/or draft storylines of this series idea, it's clear that though she has questions to ask, she's only prepared to let minimal details leak out about the tentative project she researches. Taking my cue from James Bond films, I volunteer to eat any evidence, "If I tell you any more then I'll have to shoot you". That said, without previous experience of speedway, Hester implicitly reveals – through her thumbnail sketches of the cardboard cut-out story arcs and the (cartoon-like) characters – the lowly status speedway must hold in the vivid imaginations of the editors and writers for this prospective unnamed Channel 4 crime programme strand. Questions Hester asks include:
Could someone disappear into the speedway community?
Where do they live?
What cars do they drive?
How does the community react to outsiders?
How does the community react to unusual behaviour?
Is speedway a full-time job?
What else do they do?
From what you're saying, they travel huge distances and, sometimes, they sleep in their vans – how can they afford to do that?
What does it cost to be a speedway rider?
Are there many disputes?
Serious disputes?
What do they get paid?
What do you think is the attitude towards the police?
And rules?
How do disputes get resolved?
Is there a drink or drugs problem?
Groupies?
What age are they?
Where do they come from?

What would they do if they weren't speedway riders?
Does the community sort out any problems itself or does it involve outside bodies like the police?

Hester asks way too many questions. It's quickly apparent my books don't contain enough salacious information. Though sadly they're not on sale in Andy Griggs's trackshop, I tell Hester for further research into shadier goings on in the speedway world, then almost any issue of Backtrack has something. Or, if she seeks real life stories of fights, lawlessness, match fixing, drugs, groupies and general shenanigans, then Tony MacDonald publishes a whole raft of ex-rider autobiographies in his Retro Speedway imprint. Presently stood in a world she wouldn't usually expect to enjoy, Hester's keen to offer reassurance. "I'm a researcher, I was only told yesterday. I might be in touch again, it depends how they want to flesh out the characters or develop them with this and the other information I find out. I really don't know." Hester reconfirms that my books are, effectively, pretty sanitised, "That's very unusual to send your chapters to people before you publish! I suppose you are part of the community and want to remain so." Earlier, Hester moseyed around the stadium, "They make a big thing out of the VIP area but it's only a little square!" Nonetheless, her son Eddie likes the speed, flamboyance and competitiveness of riders' racing and is surprised by their matter-of-fact reaction to crashes. "This is much better than Formula 1!"

With all the questions, it's been hard to concentrate on the meeting. After seven drawn heats (out of eight) consecutive Rockets 4-2s follow. Steve Boxall wins heat 11 and gains praise from Craig Saul, "He gives 110% effort. He's got so much potential but sometimes it doesn't go as well as he'd like." Heat 12 sees a win for Nelson, while the next race takes three attempts to run. Nine points up with two races to go requires Craig Saul to outline the permutations, "Should Rye House produce finishers in heats 14 and 15, they should win the match! The Rebels still have that outside shout". With Chris Neath first and Nelson third in the penultimate race, Rye effectively seal the deal. Ultra-professionally, Craig remains wary, "barring the catastrophe of a 5-0 that's a [3 point] win for Rye House!" Captain Chris Neath celebrates his race win with vigour, "To paraphrase the song – Don't Stop Me Now! Chris Neath is having a fine time out there on the track." Neath's enthusiasm and good form continues into the last race where he follows Jordan Frampton. "We've had many drawn results this evening and only three maximums, and one of them comes in the last race of the evening!" Celebrations continue but Craig remembers the casualties ("A Get Well message for Luke Bowen"). For Arnie, Somerset fail to show the resistance anticipated, "They weren't very rebellious!" Returnee speedway fan Dexter says, "It's the first time I've been in 25 years. I came here and Hackney from 77-85. It's changed a bit but it's great to be back!"

En route for the car park (and, possibly, more bananas), Craig Saul puts me right on my claim, "Len is doing a Jon Cook with the introductions on parade." "No Jon Cook is doing a Len!! Len's been doing it for years. You should hear Jon Cook he's phenomenal. Before I heard him I thought 'I know it could be Jon Cook's ego run wild' but he was amazing! It's all in his head without notes: just looking down the line of riders. Obviously, he knows his own but he knew all the Poole riders too. He gives the fans insight and insider knowledge – all without notes! I love working with someone who speaks his mind. You don't know what he'll say and that means everyone really listens. Stuart [Douglas] writes well and is a good guy. There are lots of really great people at Lakeside. I've been announced as the new announcer – though I tell Jon Cook I live two hours away so can't always get to them all – but he asked me to do the Test and Birmingham. He's so easy to work with. I enjoy the interviews. Robbo's only 34. I know I always think of him as an old fellow – he's handy on the mic and good to work with too. Troy is my 'go to interviewee' at Peterborough. He's got so much talent and he's really professional and focused this season!"

**21st May Rye House v. Somerset (Premier League) 54-39**

CHAPTER 4

# Leicester

## "It's like a Redcar setup but better!"

22nd May

Proper crowd

So far, the news of the 2011 speedway season is the re-emergence of the Leicester Lions at their purpose-built stadium at Beaumont Park. Constructed with amazing speed from July 2010 to March 2011, Leicester officially opened in front of a huge crowd in a challenge meeting versus Sheffield on Saturday 2nd April. The response of rival promoters, speedway media and fans is universally favourable. That said, it wouldn't be speedway without the low hum of criticism about matters ranging from the irrelevant to the minor.

The first ever Sunday meeting at Beaumont Park sees the Lions race fellow Saturday night club (albeit one with race precedence), Berwick Bandits. Pretty easy access from the M1 motorway, the rough-hewn stadium car park is at the far end of an old-fashioned industrial area. Neighbouring businesses include a supermarket, McDonald's and a swimming pool. Parked by a red-brick single-storey building, it's strange there's no perimeter fence, floodlights, or elevated grandstand. A blackboard announces this building as "The Speedway Bar!" I'm pointed to follow a new laid, almost pristine tarmac road between trees to find the stadium proper. What initially appears to be a desolate area of reclaimed land quickly opens out and ends by the light-brown sauna-style wooden fencing of the stadium perimeter.

Under clear skies, very strong gusty winds batter and noisily rattle the roofs of both home-straight grandstands seating sat either side of a tall referee's box made from breeze blocks. Each identical grandstand has deeply elevated rows of around 20 blue plastic flip-up seats split by a centre aisle made in the modern construction vernacular of tube, metal and wood. Apart from the odd seat reserved for a 'steward', the front-row seats carry the notice 'Reserved for disabled, lesser able and carers'.

My inner estate agent kicks in. My quick inventory of the stadium infrastructure at Beaumont Park spots an array of white painted metal buildings that house the speedway office, the club trackshop and – in keeping with the overall ambience – some impressively modern toilets. With over three hours to go before the tapes rise, preparations are underway at various refreshment kiosks and the bar – on wheels – is readied for action (drinks cost 10p a pint more inside the stadium than at the car park 'speedway bar'). Still pristine paving stone walkways geometrically pattern almost manicured patches of stony ground. Out on the track, members of the track curation team dress the surface with additional shale around the start-line area. Others brush the inside line of clean white painted bricks and also roughen the adjacent shale. Relatively newly sown grass on the centre green takes hold in the style of hair transplants enthusiastically advertised by ex-Test cricketers like Shane Warne and Graham Gooch.

Behind the airfence of the banked bends 1 and 2, two covered pits areas face each other. North Side Designz have painted a colourful mural on the pits exterior wall. It features broadsiding riders plus the words "Leicester Lions" picked out in team colours. Fans enjoy banked viewing areas on bends 1, 3 and 4 (separated from the airfence by a safety gap and green hard mesh wire fencing) and bricked spectator walkways throughout. The length of the back straight has a covered stand with a plaque listing the donors behind the construction of this "Leicester Speedway Supporters' Club lasting legacy stand". The Leicester Lions management team avoided the gradualism of recent new build speedway track construction and, instead, started with the full stadium package from the outset. The smooth red shale 320 metre track has narrow straights and wide sweeping bends (apparently echoing the blueprint of Blackbird Road).

Security man Shaun anticipates an even brighter future, "We're hoping from next weekend – when the football has finished – that we'll get 5% more crowds. Especially in three weeks – when the rugby finishes too – if young men are looking for something to do on a Saturday night in Leicester." Volunteers arrive in dribs and drabs. They huddle in conversation and put on their hi-vis jackets. So far – Shaun tells me – each meeting takes a little over two hours, "We've been very lucky with the weather – let's hope it continues! We usually finish at 9 or 9.15pm though it could be later with crashes. Last week we rushed it through because of the threat of rain." The sponsors' advertising banners by the pits gate indicates the club is an attractive sporting enterprise for a variety of local companies including Glenfield Television, Beaumont Taxis, The Leicester Mercury, Cooper Parry, www.nospario.com and, of course, main team sponsor Breedon Aggregates.

Inside the commentary booth cum referee's box, Leicester Lions speedway presenter Chris Popple stands by a complicated lit dashboard that looks like it could launch a space rocket. Chris demonstrates its functions to the new club announcer who debuts tonight. Familiar with the equipment, Chris's fingers skitter across the control panel in the practised manner of teenagers texting. The new announcer enjoys an impressive number of jingles (displayed on a computer screen) to choose from. "We haven't played this for the 5-1s [tune blares out over the stadium loudspeaker system] or this when JLO comes off [the snatch of another tune blares out] or this [boing noise sound] when Courtney goes through the tapes!"

When there's a gap in the training, I pass my pre-prepared typewritten sheet about my book and attendance to Chris and politely request he advise fans when convenient. "Have the powers that be approved this? If so I'll read it out." On the steep stairs down from the box, an angry man asks, "Is that you blowing paper all over the track?" I don't think so but it is. Linda Harris sympathises, "The wind is so strong it just picked up your folder and blew it everywhere!" Friendly members of the 50/50 team help snaffle up much of the paper and seven seasons' worth of detritus. Mortified by the loss of my notes (and embarrassing photos), I only half listen to Diane Cross, "I dunno where John has gone? He's supposed to be selling raffle tickets!"
[Jeff] "It must be your windiest meeting so far."
[Diane] "It's not the coldest, that was a few weeks ago. [John arrives back] Where have you been?"
[John]"Picking up this gentleman's stuff."
[Jeff] "He'll start in a mo – you can stop bollocking him."
[Diane] "I get lots of practice."

[Jeff] "Oh, are you married?"
[Diane] "Yes, for three years."

John hands back the papers found under the adjacent grandstand, "I go under there for the litter." Diane hopes for more helpers, "We want some young volunteers – we just can't find them!" Another man says, "I'll see if there's any on the centre green when I go out there later".
[Linda] "We have a new announcer today. He does the cycle speedway too! He'll be coming over from their meeting which started at 2pm."
[Jeff] "Where?"
[Linda] "Do you know Leicester?"
[Jeff] "No."
[Linda] "About three miles away in the city centre."

Every week Clean Cut Sports supremo Pete Ballinger records the race action at Beaumont Park on DVD, "We're now doing eight clubs this season. Belle Vue, Birmingham, Coventry, Leicester, Dudley, Somerset, Peterborough and I forget the other one – of course, we have two clubs at Belle Vue."
[Jeff] "You're massively expanding your empire."
[Pete] "We had to when it looked like Coventry and Peterborough wouldn't run. We had to do something!"

Leicester Lions official photographer Steve Dixon draws my attention to a photograph in the programme. "Did you see some kids set fire to one end of the pits! Tossers!" It's amazing what people vandalise. [1] Steve is also official track photographer at Scunthorpe. "The first time Plymouth came to Scunnie, Mike Bowden didn't go out on parade. I dunno if you've noticed that he always wears white socks and trousers that don't quite cover them? That day Rob – who hates going out on parade – went out with Norman Beeney and the clerk of the course Kenny Smith and they walked along together on the parade, put their hands in their pockets and lifted their trousers up to reveal their white socks. It was hilarious!" Steve's partner Debs is already on the third bend poised to send news on the progress of the meeting to the speedway updates site.

Panthers speedway fan Graham Rouse isn't sure people will spot me next to the 50/50 and programme stalls, "If they buy a programme! I have my own sheets. It saves me £100 a season at least, which is enough to pay Mildenhall!" John Jarvis and friend Bob make their first ever trip to Beaumont Park. Last season John didn't miss any Newport fixtures but my question, ("How are things at Newport?") gets short shrift, "I dunno, we haven't been going so much since Bob had a heart attack, stroke whatever it was."
[Jeff] "He must be on the mend to be here now."
[John] "It's our first time. It's about two and a half hours from Bristol."
[Jeff] "Bob looks good."

John has a tutored eye for speedway stadia since he's the expert co-author (along with Robert Bamford) of the *Homes of British Speedway*. Beaumont Park makes John's heart leap, "This is very impressive. Deluxe! I wonder if this is the sort of thing Hewlett has in mind."
[Jeff] "Is that going to happen?"
[John] "It's 50/50. They have all the money but not the planning permission from Bristol council."
[Bob] "It's the council piss arsing about again?! They were supposed to have a train in the city

[1] "I was gutted when I turned up at the stadium last Saturday morning and so was one end of the pits!! The previous night some urchin(s) had set alight to one of 4 rooms in the pits and razed it to the ground. The room was the waste oil store so you can imagine the size of the fire, also in the room were three fire extinguishers that had been emptied by the fire starters, two fire buckets, water hose and a spare empty 200 litre plastic barrel. All gone! Also damaged was the roof and the adjoining wall now has a crack in it. What a mess! The whole room had to be knocked down and the pit roof propped up with wood. I had to rush over to our extinguisher supplier for replacements and to a local hardware shop for some more buckets as these are both required under SCB rules before a meeting can take place. It's also thanks to LSSC member Dave Towers who rebuilt one wall that morning/afternoon as the Leicester City Council were due in the evening to take some noise measurements and with the wall missing the noise from the pits would have escaped. What possesses people to do such things is beyond me!"

centre and then they forgot all about it and now they've started again."

According to opinionated posters on the British Speedway Forum, the original Leicester Lions announcer – albeit well intentioned – floundered from the outset, not helped by his lack of speedway viewing experience. Things got so desperate that for last week's home meeting against Glasgow, veteran speedway presenter Mike Bennett volunteered to help him get up to speed. Sadly, this help (or, perhaps, the sheer contrast with Mike's practised but voluble style) came too late [2] Ipswich and Mildenhall speedway presenter Kevin Long bounds up, "I understand this is a refuge for washed-up presenters! Something of a graveyard!" Kevin ostentatiously checks the listing of club officials in the programme, "Oh! Announcer tba! Is it Mike Bennett?" Sadly it's not. "So last week didn't work then!"

If Mike Bennett's help with the microphone attracted criticism from posters on the British Speedway Forum so too has the shape of the Beaumont Park track. Particularly the alleged narrowness and length of the straights as well as the tricky transitions from bends two and four. Kevin Long notes, "You'd think building a track from scratch you wouldn't get it wrong? I dunno if they're limited in the amount of space but you'd think they'd say when marking it out 'Is this alright?' Let's see? It'll certainly be more exciting than Arena Essex!" Kevin spots the giant sized type of the "important notice" that reveals motorsports are dangerous and we attend at our own risk. "Oh we haven't read the important notice yet! Is it big enough? Even I can read that and I struggle with my eyes!" Witches fan and programme seller Jason is amazed at Luke Bowen's injury record, "I thought the ambulance was his van or that St John were his main sponsor!"

There's such a big queue at the trackshop doorway that fans can't get inside to buy merchandise. Keen speedway and wrestling enthusiast John Short updates me about a play on at the Brighton Festival, "It's called Big Daddy versus Giant Haystacks or something. It's on at the Old Courthouse next Sunday. It's part of the Festival. I'm interested in wrestling – if you remember – and also the theatre, so I'm going to go. No one in Big Daddy's family or Giant Haystacks's knows anything about it or the theatre company! You've missed the best thing about coming here, of course, the announcer who has left! I must find out who has replaced him."

Though speedway at Beaumont Park attracts new fans it's also brought a good crop of returnees. They need very little encouragement to visit memory lane. "I first went in 1946 when Birmingham was at the Alexander Stadium (it was also an athletics track), they moved about a bit! I saw all the riders who'd served in the Second World War: Jack Parker, Jack Duggan, Ray Duggan, lots of them. You know the other week I saw a yellow and black helmet colour. It really took me back!" Ahead of the action, Berwick Bandits supporting Edinburgh taxi driver Jim Brykajlo refuses to discuss their current number 1, "Don't talk about Charlie Gjedde. He only got 3 points last night and all but one of those was gifted!"

With the stadium thronged with an impressive attendance that the next day Pete Ballinger estimates on the British Speedway Forum at around 1,900 (compared to 2,200 for the Glasgow meeting the previous week), Reading Racers and Leicester Lions aficionado Arnie Gibbons savours the atmosphere. "You just have to stand here to feel the buzz! You don't often get that at speedway meetings. One thing that struck me is that you can hear the crowd noise as the riders head to

[2] Mike explains the background and context to his offer at length. "I mentioned a few concerns I had about the PA and the general presentation and he suggested that I should drop Alan Jones (Press Officer) an email with a few ideas that might help future meetings. I took the liberty of offering my services free of charge for one evening, to sit down with their presentation team and run through the "basics" including the expectations of the box announcer, the order of results, the times after the winner is announced etc. I made it clear that I wasn't trying to steal anyone's thunder or try and take somebody's job – living in Norwich would rule that one out (gone are the days of 6 meetings a week and 400 mile round trips!) but, due to my close ties with Leicester since way back in 1986 when I was the resident DJ at the Studio nightclub in Humberstone Gate, and the fact that my partner Tracey is a Leicester lass, I have a real desire to see the club do well. I was somewhat disappointed to receive a one line response thanking me for my constructive comments and stating "we are on it" when it comes to the presentation. Clearly, that's not the case and I just think it's sad that the underlying trend in every match report here on the forum is the lack of professionalism from the presentation team."

the bend! You know you are in a crowd rather than with 43 people standing round or with the atmosphere of watching at home in an armchair." Unlike some, Sheffield co-promoter and Tigers presenter, Dave Hoggart expresses sympathy for the recently departed Lions announcer. "He had a great voice but he was a studio broadcaster! He read things out in a matter-of-fact fashion and didn't really understand the dynamics. I used to tour for 12 years with Radio Essex and, if you put studio guys in front of a crowd of 500, they'd shit themselves! Whereas, I'd work the crowd with the two dancers."
[Jeff] "Can you dance?"
[Dave] "You're kidding!"

Neil Machin is also here, "Happy Christmas!" Talk from last night's meeting at Rye House about potential Channel 4 interest in a 4 on speedway sees Dave Hoggart's media antenna on red alert. "We could facilitate things with Channel 4 to give them what they want but also manage the image of the sport. Who was that bull-shitty thin guy Louis [Paltnoi]? He was doing Eating Dust? Yeh him, he had some stuff in there that wasn't good [for speedway's image] but it was in context. Maybe they'll want to film at Sheffield?" Always on message and promoting, Neil Machin interjects, "The fastest race track in the country."

Long-time Leicester Lions fan Steve Chilton comes every week, despite the keen disappointment of his last-minute failure to land the position of programme editor. "The first meeting was absolutely rammed, too rammed because you couldn't always see. The racing has been iffy! The dry weather has made the track bone hard. It's amazing to think we were here a year ago on May 14th for the first cut and this was just a green park. It's still a project in development, really. One of the real joys has been the old guys with their year badges. It's astonishing how much the stadium has cost. Some of the plans have had to change to meet the budget. The original intention was to have a Pavilion. It was gonna have 500 seats, corporate hospitality along with the riders' changing room and a bar. All those things have had to be split up – some of them are down the car park – and, of course, there was the cycle velodrome. When we identified those needs in 2007, there wasn't a credit crunch. It's all been done through fundraising and sponsors rather than private investors. We've had a construction industry muck tipping thing here. Lorries from various companies around the Midlands have been tipping and they've been charged for that so that has funded a lot of it. I'm still secretary of the Supporters' Club, our membership shot through the roof. It's up to 260! We've already raised a lot of money, which we've given to the riders directly as well as silencers for the Long Eaton Invaders as they use the track for the second halves. We contributed £9,000 to the project, mainly towards the covered terrace on the back straight but also the ecology report to support the planning application and towards David Hemsley's legal fees. Of course, there's the covered terrace on the back straight. That's what the Supporters Club has done! We feel we've played a significant part in the development of the club. I must get me brother's programme or he'll go mental!"

"Let us welcome our brand new announcer from the Monarchs cycle speedway, Rob Sensicall" blasts from the stadium sound system. Before the meeting gets underway, presenter Chris Popple snatches a few words with the track curator. "We've got Glyn Taylor here, how's the track?" "It's a little bit on the slick side because of the warm day and the wind problem. We put a lot of water down." Keen to manage unrealistic expectations, Chris continues rhetorically, "So if people experience a bit of dust later that's perfectly understandable, perfectly natural – isn't it Glyn?" After Glyn confirms ("It's unwanted, but yes") talk moves on to more serious matters, "We have to pay our respects to Matej Ferjan who passed away today." The crowd immaculately observe the minute's silence in honour of the 34-year-old, unexpectedly found dead in Poland (from natural causes) sat inside his car. "Thank you for that mark of respect. Let's have a round of applause, may he rest in peace!"

From high in the home-straight grandstand overlooking the first bend, I can really appreciate Beaumont Park. Everything is readied for action. Over by the pits gate there's a hubbub of bikes, mechanics and riders, the centre green is staffed and the track prepped for action. The start marshal is out by the tapes while the start-line girls are part way through their choreographed routine. This involves various nifty twirl-cum-dance moves and an air kiss gesture. The back-straight terrace is thronged. There's real depth to the crowd on the bends, plus both grandstands are full and fans line the home-straight fence. The stadium is new but the crowd size is reminiscent of an altogether different era. What looks likely to be a drawn first heat ends prematurely at the start of the last lap when Charlie Gjedde rides hard (but without contact) under Sergey Darkin. With ultra gamesmanship, Darkin ostentatiously struggles for control, travels up the sloped incline towards the fence and falls theatrically. It's a manoeuvre due marks for artistic impression. Instead, the home fans greet its completion with a fanfare of loud boos. Whoever is at the computer screen controls in charge of the stadium jingles definitely fulfils Chris Popple's pre-meeting instruction to make full use of the wide pre-recorded repertoire available, albeit inappropriately with a loud snatch of "Ha ha ha wipe out". To widespread incredulity SCB official Barbara Horley disqualifies race leader Gjedde. Sat wedged between Graham Rouse and Buxton presenter Graham Tagg, this looks a harsh decision. Graham Rouse shrugs, "He was lucky to get away with that Sergy. That's Barbara Horley innit!" With the race awarded, announcer Rob Sensicall rolls the pronunciations round his mouth, "With third place for Jade Mug-way." Both Graham's have seen every Leicester Lions home meeting. Graham Tagg also race-day announces at Buxton. "I came to the first meeting and the announcer had never been to a speedway meeting before! For the first six heats, he gave helmet colours and no times. I offered my services but I heard nothing. I feel sorry for the bloke. I went up there – the referee was Dave Watters – and said, 'You need to give the rider names and progressive scores'. I told him, 'You've got to develop your own style'. He hadn't even filled in the programme after six heats! Though people complain, he was getting better and – what made it worse – is they brought in Mike Bennett last week. I couldn't follow him!"

Despite our panoramic view the grandstand, without x-ray eyes or bionic ears we're unable to gauge Berwick Bandits reaction on the other side of the pits wall to Barbara's decision. We can easily guess, though. Graham Rouse surveys the scene proprietarily, "It's absolutely brilliant here, isn't it?" Heat 2 sees Jamie Courtney fly from the tapes in his Stars and Stripes kevlars to lead until he locks up. Clearly blasts round the outside of bends one and two require discretion, cunning and good spatial awareness. The unique topology of Beaumont Park sees the wide sweep of the bend two exit narrow and intersect sharply with the long back-straight fence. This layout catches out the Bandits number 6, Tamas Sike, who loses third because of a sudden shut off and course alteration. With commendable but unusual exactitude, Rob Sensicall confirms the race time to three decimal places at 69.016.

A rerun heat 3 sees a maximum for Berwick, Chris Popple remarks, "Berwick fans you'll make a noise now! Richard Hall is carrying some more wounds – no excuses – but I was speaking to him in the pits and he's feeling second-hand after his falls at Berwick." Richard Sweetman wins heat 4, throughout Graham Tagg urges Lee Complin on to greater achievement, "Go on Compo! Go on Lee!" New announcer Rob Sensicall continues to savour his verbal encounters with exotic and strangely named speedway riders. Though he avoids the chance to roll the absent Hynek Stichauer's name round his mouth, he still jumps into the pronunciation deep end, "We can confirm their rider replacement will be Thomas Sy-key." Channelling his inner Elton John, Chris Popple is part presenter, part weatherman, "The sun is going down, let's hope it's not too much of a problem this week!" Graham Rouse wonders, "What nationality is Sike?" My news that Sike is a Hungarian British passport holder prompts sang froid, "I don't know many of the Premier League riders because of watching Coventry and Peterborough." Celebrating his own detective skills, Chris

Popple proudly says, "We found a Berwick fan in the crowd who comes from Leicester!" Unaware that the Brykajlo family are here from Penicuik outside Edinburgh, Grahame Rouse sighs, "We don't see so much away support nowadays."

Contrary to agonised reports from Berwick fans on the British Speedway Forum, Charlie Gjedde again looks extremely fast and impresses with a comfortably easy heat 5 race win. Graham Rouse believes experience is a key success factor for first-time visitors, "It took Gjedde about one lap to weigh this track up! Jason Lyons and Gary Havelock looked good here." In addition to his weekly visits to Leicester and Buxton, Graham Tagg watches speedway elsewhere. "I've only been to Stoke once this season for their meeting with Sheffield/Scunthorpe. They took an hour to run the first race after the start gate got broken and it took them an hour to decide what to do! They've dug up the greyhound track and piled up the dirt by the fences and put in an Amco fence for the stockcars." After two seasons of greater National League glory, indications are Buxton's run of success might come to a sudden halt. "There's not many riders who want to ride for £10 a point! There's Morris and Bradford – who doubles for Redcar – they were lucky to keep Cook last year. Belle Vue signed Garrity and, then, they decided to run a NL team – so, Buxton can't use him. At reserve, there's this Ryan Blacklock. He's about 20. He's not doing bad. He's only been going round on his own previously."

The crowd reaction before, during and after each race is notable and high volume. Though fans at tracks throughout the country regularly dutifully applaud, Beaumont Park reactions to the cut and thrust of each race sweeps you along with the general tension, anticipation and enjoyment! Lee Complin's cutback on bend two and blast up the back straight to pass Sergey Darkin is greeted with a collective intake of breath and solitary yelp. Graham Tagg is sardonic, "She'll probably exclude him for overtaking him!" Rob Sensicall handles the post-race results with increasing aplomb and a hint of overconfidence, "Jamie Oliver, John Oliver I should say!" Pranksters collude to trip him up, "We've got a birthday for Joe King here [pause] we'll blame the security guard for that, he brought it up to me!"

A "break for the sun to go down" is taken at around 8 o'clock. Though Peterborough Panthers track Nicki Pedersen this season, Graham Rouse doesn't expect Elite League success, "I think we'll struggle to get 40 points. I can't see us more than mid table when we've only had four meetings so far!" Ten-minute delay over, we see a hard first corner from Richard Sweetman. It belies Chris Popple's,"sweet like chocolate" description. Sweetman's win prompts a brief blast of 'Waltzing Matilda'. After the heat 7 race results, Rob Sensicall continues, "Happy Birthday to Anthea Stanley who's 43 on 14th May. It says ('don't mention her age') but it's in brackets. I'm trying to decipher the handwriting, maybe, it's the 19th?" The all-action Kozza Smith sweeps past Jan Graversen on to win and, with Jade Mudgway third, the scores are level. Graham Tagg savours the spectacle, "There's been quite a bit of passing here after the first few weeks. They've got a bit more water into it and it's more grippy!" Though new to Beaumont Park, experienced Lee Complin uses track contours and typology to block a heat 9 outside blast from Richard Hall by cannily running wide to the fence.

During the interval, Kevin Long is almost breathlessly appreciative of the view. "It's got great viewing over there on bend three – you can see the whites of their eyes as they come down the back straight! Given what they've achieved so far, they can really build it up further when the economic situation improves!" Ipswich and Edinburgh fan Mark Shrimpton is also impressed, "I was here for Redcar, there was a lot of good clashes but fewer passes." Steve Chilton is breezy, "When Sergie fell off, I said to my brother 'as Barbara Horley's here tonight I think we could get this one!' A few weeks ago, he did that against Rye House when no one touched him and also got the decision!"

With the meeting evenly poised, the teams exchange heat advantages. Graham Tagg believes the Lions would do better if they explore all their options, "I would've put Graversen in for Oliver. Both teams' reserves are weak though, I think, Berwick could shade it. It all depends what happens in heat 14." Kozza Smith easily wins a drawn heat 12 but it's the fall and remount for Jade Mudgway as well as blistering pace of Mathieu Trésarrieu that catches the eye.

The winner of tonight's raffle prize donates the £185.50 straight back to the club (via the club mascot). While we wait for heat 13, Graham Tagg grumbles about the revolutionary second half of the Premier League season. "They're on about cutting costs but what happens if all the Thursday night tracks are in the same section? It should have been a north/south split. The problem is there'll be a lot of juggling to get in the right section. On the BSPA website it says the structure is one thing whereas in the rulebook it's down as 'odd' and 'even'". After the race, Graham continues, "He's built a stadium and found a team whereas most build a team and go from there! They deserve to succeed. It's great, they've done a fantastic job!" Richard Hall comes out for heat 14 ("he only knows one way – round the outside no matter what the track is!") and things look good for the Lions right up until Josef Franc gates to win. Richard Hall occupies himself throughout by riding Mudgway very hard.

Berwick appear to really attack the Beaumont Park track and, arguably, deserve an away win (particularly given the decision in heat 1). However, with three Premier League home defeats already, Leicester need to deny them. Graham Tagg would suit team management, "You want your gaters out in the last here! I'd go for Franc and Complin." When I ask "Which is the best gate?" I'm told, "it'll be gate 1!" Sergey Darkin rides from gate 1 in lacklustre fashion and trails off at the rear so much so an exasperated OAP shouts, "Sergey – for fuck's sake!" The meeting is drawn and the final race time shrinks, "67.72, 65.72, 64.72!" Oblivious to the cold, dressed in a skin-tight England rugby top, is referee Dave Watters, "This has been a good meeting – some here have been a bit dour."

Since his first speedway meeting in 1965, John Jarvis travels widely, "By the way, this is my 100th track in this country – that's actually racing! I actually thought it would be Exeter. When Bristol opens next year, it'll be 101. I've been to another 120 tracks in this country just to photograph them. I've got every book and magazine back to 1928. It occupies one and a half rooms. I used to have a three-bed house, it's now a one and a half bedroom house. I've been collecting for 45 years! Some people collect programmes and badges but I believe the history of the sport is in the magazines!" Some Berwick fans are unhappy. "I'm a male chauvinist pig. I've told her to her face. Women shouldn't be allowed to referee. It's one of the clearest sightlines in speedway and, if there's gonna be a decision like that, they might as well not have refs!" Steven Brykajlo is much more equable, "If you take the philosophical outlook, you'd take the draw if you're expecting a gubbing! They've got a nice stadium and floodlight. It's like a Redcar setup but better!" With their work over, the start-line girls walk away together. The girl with the red crop-top and red hair tucks into her burger with gusto.

In the speedway office, I thank/congratulate promoter David Hemsley and the Lions PR and press officer Alan Jones, "it's how speedway should be!" You often don't realise how careworn or how, over time, you can take the speedway experience for granted. After this experience, I can't wait to watch another speedway race – anywhere! My next visit to Beaumont Park also can't come soon enough.

**22nd May Leicester v. Berwick (Premier League) 45-45**

## CHAPTER 5

# Kings Lynn

## "Looks a little grip. I must make good track. Good race!"

25th May

I love speedway model

During the close season Elite League shenanigans King's Lynn stepped up from the Premier League. Almost inevitably, the King's Lynn Stars confound their critics to become the surprise team of the 2011 campaign (so far). Fifteen minutes before the turnstiles open, there's an impressive queue to pay or enter the stadium. Two riders debut this evening. Australian Sam Masters for the Stars and Russia's gifted Emil Sayfutdinov for the Bees. SGP rider Emil is box office. His signing instantly rejuvenates the Coventry title challenge and gives further overall impetuous to the 2011 Elite League. Sadly, tonight Coventry have a slight make-do-and-mend line up with guest Troy Batchelor in place of Przemyslaw Pawlicki and rider replacement for Kenni Larsen who's stuck in Denmark because of the ash cloud! At the Stars pit gate, I'm greeted by a fixed stare from lurking club incident reporter Kevin '100% professional' Moore that's more frightened flier on the verge of an embarrassing incontinence episode than Implacable.

Notoriously, the Chapman family continually redevelop the stadium infrastructure of the Norfolk Arena. Parked adjacent to the home-straight grandstand bar and trackshop is a mobile study centre. It's impossible to miss but only attracts a few visitors brave enough to step inside and meet "the training specialists". After various incarnations over recent seasons, the 2011 vintage of the programme is traditionally priced (£2), sized, and colour throughout. SCB official Chris Gay stops by, "I didn't know you'd be here. Does it say that on your itinerary?" Colin Pratt interrupts our conversation to demand, "Can I have two minutes please?"

Keith 'Buster' Chapman bats off my congratulation on their EL elevation. "Well, we'll see how it goes at the end of the year, eh? That's all you can do." Buster's wife Cheryl works hard at the turnstiles while Jonathan Chapman meets and greets new arrivals. The "Sultan of Cheese" aka Mike Bennett, the King's Lynn Stars experienced but modest presenter arrives. Jonathan updates him, "There's an ice speedway rider here tonight trying to break the track record, like he has at Sheffield and that. Dad didn't tell me so it's last minute notice. Emil Sayfutdinov is here but his English isn't very good but, if you want to speak to him, I can get a translator! Thanks to the volcanic ash cloud Kenni

Larsen is stuck in Denmark. At number 2, it should have been rider replacement but now it's Troy Batchelor so that should make it interesting!" After the sometimes quarter-witted Mike Bennett goes off, Jonathan stops by my table. "It was meant to happen! We got the chance and seized it! We got it altogether in seven days. We chatted to all our sponsors before going ahead! And then, within seven days, we had our complete team. Not many people fancied it back then but look at it now! We decided to get our team straight away. Look at what happened to Birmingham, everyone said you can have this person and that one but they didn't speak to them and things turned out different! Everything has gone great here, so far!" King's Lynn even recently won at Eastbourne. "It was great to go there and get the win back. They were so lucky to win here. We lost Olly Allen in the first race and the reserves weren't quite up to it. And then we fought our way back into it and were level going into 15, then Kenneth had an engine failure on the third lap (when we were on the 5-1). He nearly took Mads out and then let them through! Mads still fought back to second but we were unlucky. Or, put it this way, if there was any luck – we didn't get it! Eastbourne were lucky to win at Lakeside too, if you remember. Still it's good to see Eastbourne top rather than Poole!"

Without warning, the stadium speakers flicker into rude life. Mike Bennett spews forth his trademark lava-flow of platitudes mixed with exaggerated self-regard. "Welcome to a very special night here at King's Lynn, it's not just that I'm here after about six weeks away! It's not all good news, the ash cloud has done its worst, so we believe, one rider isn't here but we will get the referee to confirm that later and, of course, a very special welcome to all you Coventry fans who haven't been here for a while! We're all looking forward to tonight's racing. As usual we'll take dedications and shout-outs. You can contact me @mikebspeedway on Twitter. In the programme you can see the name of Emil Sayfutdinov – it's a huge, huge coup for Coventry and for British speedway! I cannot wait to see this man and Tomasz Gollob here in the World Team Cup!" Mike continues at length, making exciting news sound like the small print of diarrohea relief medication.

With an hour to go before tapes up, spits of rain are as welcome as a last lap Bennett commentary. The great and the good of British speedway are in town tonight to see Emil including SCB track inspector Colin Meredith (designer of the revolutionary safety fence now used at the Norfolk Arena), ex Miss Fina Invader Wendy Jedrzejakski, Sheffield team manager Eric Boocock and, out in the car park dashing towards the pits gate entrance, SCB official Barbara Horley. "Are you following me around? First it was Mildenhall, then Leicester and now here. Will you be in Edinburgh and Workington this weekend?" As usual at the trackshop, Johnny Barber and Mike Moseley are in situ. Mike wonders, "Apart from the expert glare, did Kevin Moore say anything to you?"
[Jeff] "No."
[Mike] "I expect Mike Bennett will glare too!"
[Johnny Barber] "Kevin Moore came in and asked 'have you got that book you keep promoting on the British Speedway Forum?'"
[Jeff] "Did he want to buy it?"
[Johnny] "Nah, he was just being funny."

Apparently unaware that some King's Lynn riders publicly question why they have to speak to him, Mike Bennett whimsies away on the speakers. "If you don't chat to the riders at the right time, you can't get them! If I don't chat to Olly Allen now he'll be asleep in 20 minutes because he's become a new dad ... so young too!" Like proud new fathers everywhere, Olly needs no second invitation to talk fatherhood and babies, "As soon as I found out she was pregnant things changed! You're not quite so selfish." Olly isn't fazed by the idea of Emil, "He might be crap on our tracks because they're so small!" Rather than use the translator available, Mike Bennett decides to spare the crowd any insight from his interview by wading in. "Emil's English isn't so good so we can always interview James Brundle but, then again, maybe not! Your English isn't very good so I won't ask many questions. You must be very excited to be here in British speedway?"

[Emil] "Thank you very much."
[Mike] "What are your thoughts?"
[Emil] "Yeh, I don't know what you say! First meeting for me."
[Mike] "Have you seen the track?"
[Emil] "Looks a little grip. I must make good track. Good race!"

With his world exclusive of the first British speedway interview with Emil over without barely touching the sides, Mike Bennett turns his attentions elsewhere, "Troy, always shy, always stuck for words, let's reverse that tonight. I'd like to say it's good to see you but I can't!" Troy chunters on but soon concludes, "Colin Pratt has brought me in for Coventry for a couple of hours so we're going to go for the win."
[Mike] "Obviously!"
[Troy] "I'm trying and going to keep progressing."

To relief in the pits, Mike Bennett temporarily stays his rider interviews in favour of stream of consciousness chatter. Statistically Mike must utter something cogent or relevant, "It says in the programme the next meeting is June 8th, it's not, it's next week – it's Ash's meeting. Ashley Jones's [Memorial] meeting!" With his pre-race duties completed, SCB official Chris Gay contemplates life at the base of the referee's box stairs. "I've finished all my paperwork. I love meetings like this, just feeling the atmosphere! It's lovely to be a real part of something like this. I really enjoy being a referee. Though there are some things that aren't so much fun – like making decisions people don't agree with, but that is the job! The only drawbacks are the driving and not being able to support a team."
[Jeff] "You have to be outwardly impartial."
[Chris] "Outwardly and inwardly! Imagine if you did support a team and your rider got a flier in heat 15. That wouldn't be right. I do miss being able to support a team. I always ask trainees if they are prepared for the driving and no longer being able to support a team. It is a big loss but being part of meeting – especially exciting meetings like this – are something special."
[Jeff] "Lucky you're so professional."
[Chris] "You have to project an image – your authority as the referee – and apply the rules fairly as you see them. Whatever others say or think, that's the job! This is a big crowd but they've already had a few big crowds here – the Good Friday one was the biggest in 20 years!"

If anybody sustains a head injury, experts advise – while you wait for medical help – you should talk continuously to try to stop their loss of consciousness. Despite the medical staff on duty, Mike Bennett obviously believes we've collectively suffered a severe head injury. Actually, come to think of it, prolonged exposure to such piffle could make you wish for some ailment that (temporarily) prompts a loss of consciousness. Possibly grinding our teeth in unison, the crowd suffer in silence the maestro's tried and tested puns, "it's rider replacement – that famous Norwegian rider!" Stunned silent by his wit and repartee, Mike exhorts the crowd to react at his command. Seals perform tricks in the full knowledge that they'll get fish. Quite what our motivation is continues to fox gifted minds. The standard weekly reaction of the King's Lynn faithful is, of course, resigned studious indifference. "When the boys do come out on parade in five minutes – as very much noise as you can muster up for them would be much appreciated." Instead, we greet them with a sustained initial burst of silence, "How about a bit of encouragement for the riders as they come out on parade? Come on! [hint of wayward applause] Have I gone deaf?"

Ipswich and Mildenhall presenter Kevin Long is excited. "It's amazing how one person can fill a speedway stadium. I wouldn't have come here but for Sayfutdinov." Ryan Fisher is the pantomime villain of the Coventry team. Mike Bennett briefly provokes some low level boos with a theatrical introduction. Gleeful to prompt any sort of response – no matter how muted – Mike exaggerates, "You could hear that in Swaffham!" In addition to the thrill of the meeting and the debutants

Emil and Sam, Mike promises us, "We'll have a chat with Alun Rossiter in a minute – another one who's never lost for words. We'll be asking him how come Coventry got to sign Emil." Injecting welcome normality, club announcer Edwin Overland sounds genuinely enthusiastic. "Tonight does really underline why the Stars stepped up to the Elite League. They wanted another level, another challenge and that is here tonight with the defending champions!" Gawping at his phone on the centre green, in a hushed voice that suggests the likely arrival of earth-shattering news, Mike tells us, "Tweets are coming through now." Before we learn more, Edwin interrupts to admonish careless parkers, "I have details of the three vehicles that the police have requested be moved so you would be advised to do so!" Vehicle chatter about a silver Vauxhall Astra, blue Vauxhall Sapphire and green Transit van provide welcome respite from our forced imprisonment inside Bennett-world.

When the first race gets underway, it's a thrill to actually see Emil in action. Sadly, from a neutral's point of view, Emil fails to gate. Though he fights his way back up through the field, he's unable to get close to, let alone catch, the Stars diminutive Danish number 1, Kenneth Bjerre. That said, Emil does attack the Norfolk Arena track in his own inimitable albeit slightly unusual manner. Emil enters each bend at speeds beyond those normally advisable. Not only does he remain in full control but, somehow, retains sufficient balance to hug the inside line with noticeable aggression and velocity.

The post-modern soundscape artist in Mike refuses to die so – as is his wont – he fires his microphone into life throughout the duration of the last lap. Mike labours under the delusion that his job description requires that he talks the riders round to the chequered flag in a tone normally reserved for would-be suicides perched precariously on a ledge. An unfortunate side effect of these regular interruptions is how it detracts from our collective enjoyment. It also ensures that the roar and noise of the bikes booms – albeit with a slight time delay – like a ghost echo over the stadium loudspeaker system. Confirming that he holds no misapprehension about our partial sightedness, Mike continues to instruct us on how to watch and react, "You're not going to be able to take your eyes off this meeting tonight! Auch, they've gone back the short way to the pits not allowing you to cheer!" Kenneth Bjerre won in 59.04 seconds but it's the Russian import we hear about more, "Emil picking his way through the traffic impressively."

Coventry Bees press officer Dave Rowe deserts the pits to walk the length of the home straight prior to heat 2. "I've left the pits because – after the first lap – you couldn't see anything for dust! Emil didn't make the best start there but fought his way back."
[Jeff] "Kenneth Bjerre is so quick and light."
[Dave] "So is Emil, he's small and slim too!"
[Jeff] "How long is he here for?"
[Dave] "For the whole season, if needed. He's rearranged a lot of his schedule so – apart from the GP's and a Russian Pairs competition – he can do everything!"
[Jeff] "Rosco must be very persuasive."
[Dave] "He is."
[Jeff] "It's very exciting, either way. I see Colin's out on the centre green keeping a close eye."
[Dave] "He always does that at away meetings, especially if we aren't gating that well!"

Colin Pratt's centre green presence works wonders when the race finally gets underway after Lasse Bjerre's initial tapes exclusion (and 15 metre penalty). Josh Auty and Richard Sweetman lead so easily they can indulge in some gentle rivalry. So much so, Mike Bennett says during his last lap commentary, "Steady on boys, you're on the same team!" Often happy to gratuitously insult visiting fans at the Norfolk Arena, Mike critiques those of a gold and black persuasion, "the Coventry fans who are here in body but not in voice!"

Edwin continues to admonish recalcitrant vehicle owners, "The request from the police was urgent, they do need moving!" After squinting at his handheld device – to filter out the best tweets from the chaff – Mike says, "Please give a shout out to Kevin Long – the speedway presenter who does Mildenhall and, I think, he does Ipswich, I don't know? I can't keep up!" Surely this is basic information? Next Mike will claim he doesn't know the name of the Cardiff Grand Prix presenter whose job he once used to so (via unsubtle hints and a Facebook campaign) covet. Perhaps it's arrogance that prevents Mike from a modicum of basic competition analysis? Delightful though his stream of consciousness presentational approach is, rather than have a research-light-going-through-the motions-phone-it-in quality, just maybe he could seek to keep up with other better presenters, even if he feels no need to emulate them? Whether this ignorance about Kevin is *faux* or real, Mike's already swept onto his next tweet message, "Why don't the King's Lynn fans like Ryan Fisher? Eh, how long have we got?"

Heat 3 is a race notable for Mads Korneliussen's blast from fourth to second down the home straight on the second lap. Auty gifts Sitera a point in heat 4 but Mike's attention is elsewhere, "I've just seen the line-up for heat 5 so I'm going to stand here with the best view in the house." Emil adds to our entertainment when he again fails to gate. In the first corner, to try to add to his woes, Olly Allen provides a massive shove to absolutely no effect before Emil powers away to win easily. On the centre green, Mike Bennett is ecstatic, "Wow! That is talent you've got to say! Alun Rossiter says he is box office – he's the best thing to come to British speedway." Edwin Overland's equally smitten, "Looking sensational there just as he does at the Grand Prix's – not so much last year – the race winner is a young man in a hurry from Russia, Emil Sayfutdinov." Diplomatic as ever, King's Lynn's star presenter opens his interview with Mads Korneliussen with, "I'm not sure who the bloke in white is, he was quick." Nonetheless, Mads remains confident of victory, "Yeh, we can do it! Lasse was unlucky in the first one – hopefully, he can cheat the ref in the next one!"

Though we don't learn their content, Mike assures us that "the tweets are flooding in! There's so much to look forward to. We'll have a chat to Rosco in the pits, he's always value – I won't say good value." Contrary to pre-meeting expectations, the fastest most eye-catching rider on display so far tonight – Kenneth Bjerre – effortlessly wins heat 6. Mike's too busy fiddling with his handheld device to really celebrate Kenneth's good form, "And another tweet is coming in: Kevin Long is getting more mentions than me tonight!" With Coventry already on a 5-1, Niels-Kristian Iversen falls painfully on the third bend of heat 7. He stays down for quite some time to receive medical attention. Kevin Long tweets Mike Bennett to ask if he's read my latest book or if he'll mention it. Mike replies sharply, "No! Thank you very much for your question and No! I won't be mentioning it either!" Johnny Barber smirks, "I think he's not talking about you." Ryan Fisher bests Lasse Bjerre in the three-rider rerun of heat 7 and, with Auty third, the Bees gain a narrow 2-point lead.

After a drawn heat 8 we learn Niels-Kristian Iversen is to withdraw "with what looks like a dislocated shoulder". He'll miss heats 11 and 13. Heat 9 features a battle royale between Mads Korneliussen and Edward Kennett who exchange positions throughout laps two, three and four. The race ends in a Stars maximum heat advantage to give them the lead. Only too happy to reveal the true nature of their respect for him, Mike Bennett's request ("and do me a favour, boys, come and go all the way round!") is ignored. Edwin Overland continues to be full of the joys of Elite League life and throws around phrases like "tremendous spectacle" or "racing from the top drawer" with casual abandon.

The interval features a Speedway Riders' Benevolent Fund collection by riders from both teams. Out on the track, Britain's number 1 Ice Speedway rider Kevin Rowand, is to attempt to best the Norfolk Arena track record. He's got a choice of two track records to beat – either Kevin Doolan's Premier League record (58.3 seconds) or Nicki Pedersen's Elite League version (57.60 seconds). Mike Bennett bigs up the potential excitement, "Kevin Rowand told me he took some time off the Ipswich track record." Mike witters on at length before and during Kevin's ride. Edwin Overland delivers results news, "Well, he hasn't beaten the track record." In fact, with a speed of 60.97 seconds, so far tonight Kenneth Bjerre (twice), Emil Sayfutdinov and Niels-Kristian Iversen all record faster times on bikes without spikes! Nonetheless, Mike Bennett continues to talk up the thrills, "around those bends, didn't he generate some speed!" He also can't resist backhanded criticism masquerading as expert analysis, "there were a couple of mistakes we spotted!" Johnny Barber prefers to serve customers rather than watch Kevin's derring-do. "After all those mentions, I was expecting some Dexy's Midnight Runners to play. The song 'Come on Eileen' would have been a highlight! The best thing is his van that says "International Ice Speedway Rider". Not being funny, it's gonna be international, because we ain't got any tracks! So you have to travel!"

Responding to the remark, "That's a shame about Niels-Kristian Iversen's injury", Stars team manager Rob Lyon is studiously matter of fact, "That's speedway, we've gotta dig deep and keep going." To further slake our desire for additional medical insight, Rob continues, "Yeh, it's dislocated shoulder. He's got it back in but it's very sore and very weak." Referee Chris Gay savours sight of Emil on his British speedway debut. "You probably can't see it from here but he dives tight and keeps the speed as well as being able to hold it – as opposed to drifting wide [gives flamboyant but, sadly, silent demonstration on his imaginary speedway bike] without any loss of speed. Holding his speed, he takes a line no one else does! This is going to be fairly close. It's an exciting meeting."
[ Jeff ] "I think King's Lynn will win."
[ Chris] "Really? What without Niels-Kristian Iversen? Heat 11 is a 1-5 and 13 is a problem too! Rosco was telling me that two people have come from Huddersfield to see Emil and they're not even Coventry fans! I was disappointed not to see Nicki [Pedersen] on the telly on Monday. It was a real pleasure to see him ride here at the height of his abilities. He's not quite what he was and has, probably, achieved all that he set out to do."

Edwin Overland draws the interval SRBF collection to a close, "and there is a call for the riders to return to the pits from their collection – it's not often I get to tell them what to do!" Coventry physio Jodie Lowry's dad Russell tries to describe the unique trajectory Emil Sayfutdinov shows around the Norfolk Arena, "It's almost like a diamond thing like Gollob does." Russell's friend interjects, "He's only little, he looks like a 12-year old! He's strong and confident on the bike though." Ipswich fan Jason is also mesmerised by his flair. "I've got to say I think he's the most exciting rider I've ever seen! When we had Gollob that was brilliant but he's a cut above, particularly the way he rides the bends." Jason mimes a tight turn with his own set of imaginary handlebars, "They're very small handlebars!"

Heat 10 only lasts until lap two when Josh Auty runs out of room on the transition from bends three to four, hits the airfence and flies over his handlebars to take what Edwin Overland describes euphemistically as "an alarming-looking tumble". Heat 11 finds Emil back out on track. Departing from our collective script, Lasse Bjerre pressures him for the first lap. Troy Batchelor isn't the best choice of Coventry guest tonight, so much so that Filip Sitera (riding in place of the injured Niels-Kristian Iversen) bests him for third. Mike Bennett finally manages to pigeonhole Rosco and wonders aloud (again) about the Emil signing, "How did you do it?" With all Mike's build-up to this supposedly revelatory interview you'd half expect to hear a tale of drama, intrigue and cunning. However, the reality is much more prosaic. "I just picked the phone up and took a chance. I've always been an admirer. I just rang Suzi [his manager] and he said it was 70-30. He rang back half an hour later and I sent fixtures. He rang another hour later and I thought he's keen!"
[Mike] "You have a bit of a relationship with the [British Speedway] Forum jockeys."
[Rosco] "Yeh, they said he's not coming, he's not gonna do all the fixtures." They're idiots! What do they know?"

[Mike] "He hasn't seen it before and most riders take a couple of races to get used to the track but he's been right there."
[Rosco] "To be fair, he actually rode the Norfolk Arena in the World Team Cup. He's asking a lot of questions. His set-up is fantastic. His bikes are unbelievable! He said it's incredible how welcome everyone has made him feel, even the King's Lynn fans. He says in Poland no one smiles."

Back in the trackshop, Johnny Barber wonders why Mike Bennett ignores Kenneth Bjerre tonight, "Don't worry about your own rider who's gone through the card, just kneel at the feet of Sayfutdinov and lick his arse!" Heat 13 sees the battle of the Number 1s, result in yet another race win for Kenneth Bjerre (four rides, four wins). Belatedly Mike Bennett wakes up to his performance, "Now tell me the truth – you didn't have that marked in your programmes before it started!" After some Fisher resistance, a King's Lynn heat 14 maximum seals the result. Before the final race, Jonathan Chapman is ebullient. "Yeh, I mean, when we lost Niels I thought 'Oh no!' but we've dug deep! The heat for me was heat 11 when Sitera and Lasse Bjerre went under Batchelor and finished behind Sayfutdinov to make that 5-1 a drawn heat. I can't wait to see some more of Emil. Thank you to Coventry for letting him make his debut here. A fantastic crowd has turned up! You don't usually talk about it but – for the [SRBF] collection everyone has been extremely generous – the most I've ever seen collected. There's a competition amongst the promoters to see who's collected the most at the end of the season I think we'll be up there!"
[Mike] "If not, you'll have to make it up."
[Jonathan] "Yeh, we certainly will."
[Mike] "Next week we have the Ashley Jones Memorial meeting. Can you tell us something about that?"
[Jonathan] "I've had riders ringing all the time asking if they can be involved! There is a prize money, but none of the riders asked how much it is!"

While the result of the meeting is beyond doubt, Coventry could still snatch a league point. Kenneth Bjerre has other ideas since he blitzes the Bees Sayfutdinov and Kennett to complete his five race maximum. In his personal duel with Sayfutdinov, it's three races for three wins. It's a night where the calibre of the riders and racing double-underlines why the King's Lynn Stars management team went Elite. The stadium infrastructure continues to improve and develop along with industry-wide recognition that will see the club again stage a World Team Cup round this summer.

Prior to the victory parade, Kevin Rowand makes his second attempt to break the track record. Mike Bennett isn't quite sure, "All I can say, it's different! He looked steadier that time. Kevin Rowand is one of only two British Ice Speedway riders left! Dear, oh dear!" Edwin Overland confirms Kevin again fails to set a new track record, "In a time of 60.77 compared with 60.97 last time." From the parade truck, Kenneth Bjerre is studiously modest, "I think today was best team performance in a team I begin to join."

Post-meeting practice for Stefan Nielsen ends almost immediately with a horrific looking crash on the back straight that sees him ferried to the hospital by ambulance with blue lights flashing but claxon on silent. [1] Even the spectators are in the wars. A female Coventry fan with a cut on her brow tries to leave the stadium but gets sent back down to the pits to receive treatment. Out in the car park, Barbara Horley departs excited. "Adam Roynon and I both turned to each other after heat 1 and said 'he's rubbish!' and, then, 'he's something!' I'm looking forward to reffing him at Eastbourne in July!"

**25th May King's Lynn v. Coventry (Elite League) 50-48**

[1] Though it looks awful, subsequent reports reveal concussion but no broken bones.]

# CHAPTER 6

# Somerset

## "I thought they'd travel today but there's none here!"

27th May

Rebel rooster

The proximity of Somerset speedway's impressive Oak Tree Arena to the M5 motorway doesn't alleviate bank holiday weekend traffic snarl-ups on the A38. After a seven hour 195 mile trip, I'm pleased to arrive to see the first ever meeting of Somerset 'Dickies' Rebels versus Leicester 'Breedon Aggregate' Lions. Affable club promoter Miss Deborah Hancock warns, "We're a bit depleted! We've got Doyley – he's a good guest. We love Doyley! And we've got Claus so we should be alright. Two of our lads were supposed to be on Under-21 team championship duty – Sam Masters and Dakota North. Sam has gone but Dakota hasn't because he broke his finger last night riding for Peterborough."

By the start line, Tiny Tim has exciting news, "We'll be able to see Rooster!"
[Jeff] "What?"
[Tim] "Rebel Rooster is making his debut. He's been having a suit made. It's a big suit."
[Jeff] "It would be if you are in it."
[Tim] "It looks good, wait till you see it!"

Also dressed for the occasion is Somerset mascot 10-year-old Henry Atkins immaculate in pristine kevlars sat aboard a gleaming bike. "Do you want to take my photograph?"
[Jeff] "I do in a minute."
[Henry] "Am I in the book?"
[Jeff] "You are! I spelt Somerset wrong in that chapter."
[Henry] "That's alright."

Somerset Rebels Equipment Fund (SREF) organiser Di Phillips bustles about the stadium selling tickets for the club trip to the Cardiff Speedway Grand Prix. "We'll take two coaches full again – about 100, I'd say." The well-travelled, knowledgeable Reading Racers club historian Arnie Gibbons is here too. "The Isle of Wight has an email contact address. Tim Helm encourages you to send him messages and I was tempted to email and say, 'Rather than tell us jokes that you think are funny but no one else does, why don't you try telling us who is in the next race?'" The stadium speaker system crackles into life with the voice of presenter Nigel Thomas, "Di has some travel news." Arnie doesn't need telling, "The M5 is fucked."

With a strong wind, it feels unseasonably cold though the sun has yet to set. Snug and warm in the referee's box, announcer Jordan Satchell has exciting news, "There's a special guest on the parade tonight." At that exact moment Di Phillips says, "There's a special guest on the parade tonight!" Nigel Thomas interrupts Jordan, "Rebel Rooster told me he's going to be milling about the stadium. I said 'surely that should be milleting about the stadium?'" Reporter Dave Thompson lingers by my display table, "I like Mike Bowden's centre green demeanour – it's sort of presenting by numbers, like painting by numbers, but without the paint!" Sight of the riders on parade touches the serious side of Nigel Thomas, "Over the years we've lost some of the riders through racing accidents but, unfortunately, over the last few years we've also lost some riders through non-racing accidents and that has happened with Matej Ferjan. So we'd like you to stand in a minute's silence." After the strict observance of these memorial moments, Nigel continues with rider introductions. While he does so, away on the far side of the track, Henry Atkins's bike grinds to a halt. Helpfully Nigel Thomas breaks off, "Mike, your boy's broke down!" Dedicated Peterborough Panthers fans Steve and Sarah Miles have come to the Oak Tree Arena to celebrate Steve's birthday. When I ask "Is it 41 today?"
[Steve] "Did she tell you?"
[Jeff] "She did."
[Sarah] "I didn't! Have you been on Facebook? Were you at Peterborough last night then? Or have you just been standing behind me again?"

Friendly Somerset Rebels owner Bill Hancock meets and greets many fans as he wanders round his stadium, "I reckon this is going to be a close meeting."
[Jeff] "I brought the traffic with me."
[Bill] "That's normal down here – especially on a Friday!"

Rebel Rooster passes and elicits Bill's sympathy, "I reckon it's going to be hot in that suit!" Out on the track, Jason Doyle (in place of the injured Travis McGowan) treats the Leicester pair of Serge Darkin and John Oliver to a hot racing manoeuvre. Claus Vissing soon also cuts across them imperiously. Steve Miles sighs, "This is going to be like watching Peterborough away! I already feel sorry for them and it's only heat 1!" The second race of the night takes two attempts to complete. Prior to the race, Nigel Thomas requests the crowd give Rebels reserve Alex Davies "a big cheer" to welcome him in his first OTA race. Afterwards Nigel strikes a more morose tone, "As is often the case, the second bite of the cherry allows the rider who didn't make the start first time out to win!" Heat 3 sees visitors Richard Hall and Mathieu Tresarrieu hammer home a 1-5 to level the scores. Ostensibly over his broken rib injury, Christian Hefenbrock eases back into his saddle with a third place ("maybe feeling his way back into the meeting").

After a bright start to the season – initially driven by Dakota North's form at reserve - Somerset's season is now becalmed by a series of rider injuries. The way of the modern speedway world is for riders to double up and double down as well as ride in Europe so – with these tightly packed schedules – it's statistically likely that bumps, bruises or worse will alter team line-ups. Nigel requests the injury latest from promoter Debbie Hancock, "I don't think you were up at the hospital last night after being there so much?" Speedway promotion inevitably grants an ongoing medical education, albeit by osmosis. "I had the night off from the hospital but, unfortunately, I was on the phone with Dak [North] after Peterborough last night. Dak came out of the dirt and went into the fence. He's broken a forefinger in two places and also got a broken thumb. He might get referred to another hospital and might have it pinned. Obviously, it's disappointing for us and the fans only to have six riders on Tuesday for our Sky meeting against Sheffield but we shall overcome the Sheffield Tigers, I'm sure of that!"

Nigel comes over all technical, "The tapes are blowing about in the wind a little bit this evening which may account for the riders being a little bit anxious to get away from the start." Lions reserve Jan Graversen again demonstrates a liking for the Oak Tree Arena track in heat 4 until he suddenly halts with a broken chain half a lap from victory. This transforms a possible drawn heat into a home advantage. John Oliver finishes second to make it 3 points from two rides. I tell Steve Miles, "He's already scored more than I'd expected!" It's a surprise for Steve too, "He's not that strong at home either, is he?" The OTA curatorial team briefly get to play the speedway equivalent of hunt the thimble, "This is a message for the track staff to look out for the primary chain – especially on the back straight – oh – it looks like they've found it!"

Lured into the Rebels trackshop by its comparative warmth, Andy Griggs's disappointed at how few Lions fans are here, "I thought they'd travel today but there's none here! How was Leicester when you went?" If I speak the language of atmosphere ("It was great"), Andy speaks an altogether more commercial vocabulary, "No, I meant sales!" Andy's quick to dampen my anticipation and excitement about the Premier League wooden spoon clash at Hayley Stadium tomorrow night between Newport and Plymouth. Well versed in the wiles of the speedway promotional world – though sunny spells and showers are forecast – Andy strongly suspects that this meeting will suffer an early postponement. "I don't know if it will but there's also the rugby final in the afternoon, the GP and then you've also got Manchester United versus Barcelona in the Champions League final to compete with!" Leicester fall further behind in heat 5. It's the third race of the first five to be rerun. Andy Griggs asks, "Is Ronnie Allan the ref tonight?" then pauses for effect, "You can tell because every little thing he stops the race and all the riders look at each other wondering what happened!"

Though we all know speedway is a dangerous sport, some incidents really shock. Last night at the East of England Showground stood out for Andy: "The crash last night was really bad. Ale Dryml fell and Buczkowski had nowhere to go and hit him on the head with his engine and you know how fast they come out of bend two! The whole stadium there was deadly silent. You could hear a pin drop. When I left, he was still in the ambulance, which was very strange. Steve Miles told me today he smashed his helmet but he's got a broken shoulder, three broken ribs, a punctured lung and severe concussion!" An ear-wigging fan in the trackshop interrupts this horrific catalogue to ask, "How's Nicki going at Peterborough?" Nicki's definitely fallen short of Andy's expectations, "Crap – another heat 1 loss last night!"

Last night heat 14 took time to conclude at the East of England Showground, "It was one of those silly races cos Buczkowski had to go round on his own to get 3-0. It was freezing at Peterborough last night too. It was seven degrees when I left." Out on the track, Claus Vissing wins heat 6 and with Jason Doyle second it extends the Rebels lead to 10 points. Andy's mind is elsewhere, "How many times has he announced the trackshop so far?" It's a teasing question, "I dunno? None?" "Correct. No matter how many times you tell him about the trackshop and the speedway merchandise, he forgets!" Mat Tresarrieu is disqualified from heat 7 under the two-minute time allowance. Ironically Tresarrieu was first out at the start line but, when he backed away from the tapes to establish the whereabouts of his teammate Richard Hall, fell foul of the referee.

En route to the clubhouse for a warming cuppa, Leicester Lions PR and press officer Alan Jones stops briefly. "The track is still bedding down. We're going to make some changes like pull the inside line in a bit more to create a better inside racing line!" There has been much praise about the friendliness of staff and fans alike at Leicester. "Yeh, the stewards are friendly and we've had praise for our stewardship. We've only been going seven weeks and none of us had done it before but we've received lots of praise and advice." Heat 8 is nearly underway. News that Jamie Courtney replaces Jan Graversen gets a deep sigh from Alan.

Apart from the second race exclusion, Mat Tresarrieu (along with Richard Hall) looks the Leicester Lions most impressive rider. Rebels guest Claus Vissing is also impressive and on a four ride paid maximum until his third in heat 10. Andy shrugs, "That's what comes of opening your mouth saying how much you like riding the place in an interview." Di Phillips never needs a first let alone a second invitation to discuss Magnus Zetterström, "You still haven't started *The Mask of Zorro*! Did you see he's beaten Nicki Pedersen three times in a meeting in Denmark?" Heat 12 sees the Tresarrieu in black and white and the scores narrow to 45-31. Before heat 13, British Ice Speedway rider Kevin Rowand tries to beat the OTA track record of 56.08 seconds held by Leigh Adams. The spectacle impresses Di Phillips, "Look how close he is to the shale." Another fan is more sceptical, "He's slower than Henry Atkins!" Di's fellow SREF fundraiser Sue Tomkinson is all ears, "I've got a lawnmower that makes more noise than that." Though Kevin does set the fastest time of the night (58.09 seconds) it'falls 1.5 seconds outside the fastest time of 2011 (so far) set by Bjarne Pedersen.

The sound of the ice speedway bike is irksome rather than noisy, but still temporarily deafens presenter Nigel Thomas, "For the people who didn't hear that, can you just recap on the result of heat 12 because I didn't hear that because the bike was on the line." Heat 13 promises to be a closely fought contest with Doyle, Gathercole and Tresarrieu all well amongst the points. It's a drawn heat notable because Cory Gathercole throws away his maximum with a last place ("When the man who hadn't been beaten all night had to take a back seat"). Many fans pose for a photograph with Rebel Rooster. I can't resist either. Sarah Miles comments, "Is it your ambition to be photographed with a bird at every speedway meeting? Nice of you to wear a hat that matches his mouth, beak I should say!" Though I missed it, Arnie tells me that club announcer Jordan did mention my book, "He sounded like he was reading a script – it didn't exactly sound sincere! He said things like 'It says here it has 276 pages.'"

Nigel Thomas grabs Alex Davies for a few words and opens with a potentially leading question, "Is it good to know you'll be here for the rest of the season?" Instead, Alex fails to sound off about his sudden departure from Newport, "It is unfortunate that I had to be dropped from Newport but it is a numbers game. I'd like to thank them for last season and this!" Richard Hall wins the penultimate race of the night and, with the combative Lions reserve Jan Graversen third, the score is 50-38 with one race to go. With a paid 14 from five rides in the bag, Rebels Claus Vissing comes unstuck on the third bend to earn a disqualification. Richard Hall impressively puts his bike down with great alacrity to avoid the fallen Dane.

Like Weymouth Wildcats fans everywhere, volunteer track staff member Ian Lusher can no longer attend Radipole Lane to get his speedway fix. Instead he comes to the Oak Tree Arena or travels internationally. "I always promised myself I'd do these tours and now I can! I always use Travel Plus Tours – they're excellent. The first one I did was Australia – not last winter but the one before – I went to Italy too." In Dorset things don't look so bright. "I can't really see us [Weymouth] making another comeback, that's a shame I'm sure they could have rescued it! They didn't like you in the end at Weymouth. She [Sam Knight] took exception to something you wrote in the book. She was rabbiting on but I tend to switch off when she's moaning. I finished up doing the pit gate with her but didn't have any financial dealings, so I didn't fall out with her or Phil [Bartlett]. I'm a person who tries to be friendly. The number of people I apologised to for Phil's behaviour last season was ridiculous. I used to tell them 'I work for Weymouth not for Phil!' He didn't pay me so we weren't gonna fall out. They acted a bit silly on occasions. They thought they had something against Luke Priest at one stage and he did brilliant for them. It was Sam who thought she had a grudge against you! Something you wrote she said. I must admit I only ever bought one of your books so I don't know."

The rerun of heat 15 ends in a win for Jason Doyle ahead of Richard Hall and Mathieu Tresarrieu. Nigel Thomas ends with an unilluminating interview with Christian Hefenbrock about his paid 10 from five races, "We don't want to keep you around too long because you're cold."

**27th May Somerset v. Leicester (Premier League) 53-41**

# CHAPTER 7

# Edinburgh

## "He'll come out of the gate like a salmon and then he'll come out of the bend like a whale"

17th June

Fuel fund shed

Various online weather forecasts for early evening in the Armadale area don't look that clever. Berwick presenter Dick Barrie kindly texts the latest information, "This from the Monarchs info line at 3.45 Friday 17th – 'Only spot of drizzle at Armadale all day. Tonight's match, definitely on. Will be tough for Scotwaste Monarchs with Sheffield stronger without Ashworth.'" You have to admire such a categorical message, particularly one nuanced enough to pre-identify plausible excuse just in case Edinburgh fail to win. I suspect it's the handiwork of speedway's most elegant match reporter, Mike Hunter.

Ninety minutes prior tapes up, there's intermittent light drizzle in Edinburgh city centre. In Waterloo Place, the hard core of Monarchs fans gather to wait for the Supporters' Club coach to ferry them to Armadale. Trip organiser Ella MacDonald wears a "Half Cut Crew" top with "Ella" in customised letters on the reverse. Mostly it's regulars on the coach. Some look tired from their trip to Redcar and back last night. Speedway wouldn't be speedway without overheard snatches of opinion. "Lambert was on fire last night. He got excluded in his first ride but still got 15! He could have gone through the card last night!" Dressed in hi-vis clothing, volunteer Armadale race-night medical staff member, David Nimmo is deep into a 'he said, I said, he said, I said' type conversation that only catches my attention when he says, "Then they moved the vestments!"

In the vicinity of Edinburgh airport, the sky brightens appreciably but – by the time we arrive at the M8 junction (4) exit towards Armadale – these dark clouds begin to empty. Nodding towards the steady stream of cars in the opposite direction with wipers going, our driver remarks sotto voce, "Look at all those cars with lights on coming towards us!" Our driver steadfastly resists putting on the wipers until lack of visibility ensures he chooses the lowest intermittent setting. In the Scotwaste Arena car park, the rain is more mizzle than drizzle. Despite the box of books under my arm, the bloke on the entrance gate is highly suspicious, "Yes! Can I help you?" Once inside the stadium perimeter, affable Monarchs promoter John Campbell greets me cheerily, "Do you want a table? We've got a small one that was £4.99 from Morrisons!" John soon bustles away elsewhere unaware that there's bad news at the catering kiosk ("there's a problem with the gas!").

Despite what's now drizzle, there's still a reasonable queue at the turnstiles while, at the catering kiosk, fans happily order from the more limited selection available ("sorry we haven't got any onions left"). Inside her open-fronted wooden shed adjacent to the trackshop, Margaret draws deeply on her cigarette and watches the world go by. "You picked a good night to come. Actually it hasn't been too bad." Sat next to Margaret, her friend is confident, "I think they'll start dead on tonight to make sure they finish." Margaret hopes to win tonight's raffle, "It's my turn to win it. I can't remember the last time I did. I think it was at Powderhall!" Well-travelled, cheerful speedway fan [the late] Maurice Gordon arrives in a Monarchs anorak keen to bat off my fashion garment queries about the whereabouts of his Newcastle or Berwick anoraks. Maurice sniffs, "I haven't got a Glasgow jacket yet!" Not that we quiz him but, by way of further tangential explanation, Maurice tells Dougie Copland and myself, "I'm a Berwick Supporters' Club member and season ticket holder." Dougie wonders why I didn't come up last weekend, "You missed the Super Guest Cup last Sunday!" If Dougie's speedway enthusiasm were a hot air balloon, he's developed a slow puncture. "I'm not as keen as I was! Or up to it! It's just speedway – all the same old things bother me. With this rain, you'll be here for another bad crowd. Last week was the best so far for the Super Cup. You're lucky you're not here every week for heat two when it starts." ["Why?"] Lee Dicken I'm saying no more." Dougie's no longer enamoured with the STMP either, "I'm not so keen on the Redcar track nowadays – it's only got one line now! When Chris Kerr was there, you could go up the banking."

By his own high standards, Maurice ("pronounced 'More-reece'") had a slow start to his speedway travels this season. "I'm not doing very well. I've only been to 28 meetings, so far. My record is 84 and not one overnight stay! The petrol is so expensive now, that's why I don't go to Newcastle that much. It's about £38 return – although it depends on how you drive."

At 19.25 – five minutes before the scheduled start time – Scott Wilson's voice blasts out, "All track staff take their positions please!" In the continuing drizzle, John Campbell stops by to check on my sales. At 19.29, all the riders are up at the tapes – except Sheffield Tigers guest and ex-Edinburgh Monarch, Aaron Summers. SCB referee Willie Dishington disqualifies him under the two minute time allowance. Scott Wilson expresses scepticism as to whether this will handicap the visitors, "The Sheffield reserves have been so strong this season it's probably why they're winning so many meetings!" Though Sheffield only track six riders tonight, the programme reveals an exotic range of birth locations including Adelaide, Buenos Aires, Coffs Harbor, Boston and Sheffield. Biggar-based Glasgow fan Jim Fleming will have to smoke in drizzle this evening. Jim's got many speedway facts at his fingertips, "Simon Lambert has Scottish blood, did you know that? I dunno if his dad is Scottish but his grandfather definitely is." The constant drizzle verges on – but doesn't quite become – rain. It's definitely suboptimal on riders' goggles if they race bikes with no brakes.

They're made of stern stuff in the Premier League, though Glasgow recently called off their meeting with Sheffield in mid-morning to the surprise of the Sheffield riders already at the track and the consternation of their management. It was news that also surprised Jim Fleming, "We arrived back at Edinburgh airport at 4pm that day and I couldn't believe it was off. It was a lovely sunny day." Scott Wilson's fears about the Sheffield reserves are well-founded since they romp to victory in heat 2. However, the only real talking point is the performance of Lee Dicken. He spends most of the race marooned in fourth right up until until the last bend when he suddenly rides like the clappers to beat his teammate Ashley Morris for the chip shop money. Jim Fleming no longer visits the Scotwaste Arena as regularly as other seasons, "I haven't been so much and, when I have, it hasn't been so good. It's often processional! Glasgow's a lot better now the track has settled down."

Aaron Summers actually gets to the start line (as rider replacement for Ricky Ashworth) for heat 3 on time. That's about as good as it gets for Sheffield since the Monarchs Katajisto and Tully surge to a comfortable win. The most appropriately named rider in British speedway, Hugh Skidmore, fails to complete his race after a fall. Ahead of his ride in heat 4, Scott Wilson pays tribute to long-time talismanic ex-Tiger Shane Parker who sees out the final season of his British speedway career at Sheffield. "It's his last scheduled Armadale meeting – though he might be back for the Keyline Open – Shane Parker!" Dougie Copland looks on morosely, "We've got the worst reserves in the Premier League!" Ashley Morris certainly tries wholeheartedly. So much so that, on the transition from bends one to two first time out, Simon Lambert glances behind just as the blue helmeted Ashley Morris runs into his back wheel and throws him off his bike.

Because of the weather, fans cluster under the shelter of any available roof on the home straight, bend one or the hospitality areas. Dougie asks and answers his own questions, "Have you ever seen our crowd so bad? I suppose the weather doesn't help." Something Dougie read in one of my previous books irked. "You do come out with some statements. You completely incensed me and irritated some others. It's not your fault. You're just writing what you're told. I was here. It's about four things rolled into one to just about make a false story. It could be seen as damaging until I realised the source. It was just unfortunate that you didn't bother to check it with me. I'd forgotten you were coming over or I'd have reread it. There are a number of things being said at other tracks – the management and promotion were aware – and you stepped right into it. I don't think you sent me that chapter or I would have told you!" Shane Parker wins the rerun of heat 4 with ease. Dougie bats off my suggestion, "I thought Shane would struggle in the wet" with, "The track is fine!" Aaron Summers wins a drawn heat 5 and, after the race results, the announcer says, "Would the centre green presenter please come to the referee's box?" There's immediate loose talk of a fine or reprimand for Scott Wilson among the Edinburgh fans but this soon peters out when no one can recall anything contentious or controversial.

Heat 6 sees James Grieves versus Shane Parker. It's only going to go one way according to Dougie, "Grieves should win it. Remember he's the second best rider to ever ride here apart from Peter Carr – he's better than anyone else!" Almost immediately it's effectively a non-contest won comfortably by Shane Parker ("I struck out there!") If this season's Premier League action around Armadale doesn't light Dougie's fire, there's much worse elsewhere. "The Elite League is awful. I can't even be bothered to watch it on Sky!" Since I don't have a satellite television subscription, I have questions about the newest co-commentator, "What was Jan Stæchmann like commentating on the GPs?" Apparently, you can take the man out of Denmark but not Denmark out of the man. "It's a bit biased. When Nicki had Harris off at Copenhagen, Jan Stæchmann said, 'Harris got off it there!' He mumbles like Ronnie Correy did when he did it."

Drizzle isn't good for speedway or book sales, Dougie sympathises, "Speedway fans are hard to please. There were years without books and, then, they got invaded by Tempus. If you're gonna do histories, at least do them properly!" Hugh Skidmore to race Craig Cook fails to distract Dougie from the brute fact that Lee Dicken will also ride. "Someone said to me about Lee Dicken, 'He'll come out of the gate like a salmon and then he'll come out of the bend like a whale!'" When Lee's engine stutters on the second bend, Dougie corrects himself, "Maybe like a beached whale!" This mechanical hiccup sadly relegates Lee to the rear for the duration of the race, "Mike Hunter is the only man I know who can put a positive spin on Lee Dicken!" [1]

After an uneventful drawn heat 8, the next race fires into life when Kalle Katajisto locks up on the apex of the second corner and, for an instant, looks likely to get thrown from his bike. With lightning reactions, Shane Parker almost stops and pirouettes on the spot only to see his commendable safety-first attitude enable everyone else to zoom away up the back straight. Dougie chortles, "He didn't fall off – watch him, he's going to go round with his hand up!" Shane does exactly what

[1] In the programme, someone I suspect to be Mike Hunter, writes in the Dales Tales column: "Which speedway author has written the most words? Well you might think of someone like Peter Oakes but more likely it will be our visitor tonight, Jeff Scott."

Dougie suggests. "He waits for the decision that doesn't come! Admittedly, Kalle pulled a massive locker in front of him!" His race effectively run by the first bend, Shane gestures again to referee Willie Dishington as he finishes the heat. If Shane's disillusioned with officialdom, there's a raft of small irritations that cumulatively add up to disgruntle Dougie. "I've lost my passion. It's the politics. It's the car park. It went up to £2, though it's like a minefield. It's the rain-off thing. £1 for every race or whatever it is. It's everything. It's even riders like Shane when he wants to ride it's great but – often, if he hasn't gated – he just goes through the motions!"

Shane Parker's immediately back out again on a rider replacement ride for Ricky Ashworth. Shane wins easily in steady fashion. Distracted by my programme, Dougie nudges me, "You missed Shane wave at the ref there!" Scott Wilson sportingly celebrates Parker's win, "Give it up for the rider in white!" With the drizzle heavier and the track even wetter under tyre tread, heat 11 looks much slower and processional. Dougie knows why, "Everyone is going as slow as Dicken now! Sanchez is crap, that doesn't help it either." Until the last bend, Lee Dicken looks likely to secure his second point right up until the moment Sanchez sneaks past. Won by Andrew Tully, heat 12 is the third consecutive drawn heat.

With the drizzle verging on rain everyone – except the riders, track and medical staff – shelters to watch Shane Parker's last ever programmed ride at the Scotwaste Arena. Dark skies and dampness definitely don't brighten Dougie's mood about the sport. "To me speedway doesn't deserve anything! Most of the time, it's in shit stadiums – like this – that they think 'oh speedway is something'. To be fair, the product a lot of the times is good, it's just amateurs running it. I wouldn't show your photo book to any of my friends because it shows speedway for what it is! I know it's not their fault because they don't own the stadium, but the hospitality here is in a portacabin! Why would anyone come? I live 30 miles away from here and I ask myself 'why should I come here every week?'"

Craig Cook wins heat 13. On his celebration lap, Craig holds a Saltire flag that the crowd greet with enthusiastic loud cheers. "Craig is a bit of a hero around here." There's a slight end-of era feel to Shane's last place in this race, "Is this Shane on his retirement year? I don't like what he's thinking – 'Well, I'll just watch, I'll just sit this one out.' Things like that piss me off, like he didn't even try! It looks like 'if I gate, I'm interested'". Heat 14 sees Lee Dicken lollop effectively in the first bend. Dougie notes, "He uses the old track craft there! Lee Dickens only got a point because he Sumo'd Ashley out of road!" Warming yet again to the theme of his disillusion, "Maybe I've been coming too long? We've won the League. We've finished first, second, first – I don't know what the answer is! Even last weekend the farce of Glasgow using four guests on the Sunday and Edinburgh saying 'we didn't know' doesn't help! Though they contradicted that by saying 'we tried to stop it'. People go 'Enough's enough!' What's ahead? They're trying to get the youth going. It's a sad state of affairs but there's no talent coming through! In the second half them's two guys race every week – one's from Glasgow and the other won't make it. It's not good for speedway! I've given up bringing people now as well. Generally they don't like it or, if they do, when they come the next season everything has changed with the rules and you have to say 'You know everything I told you last year, it's changed!'"

Shane Parker takes to the track again, this time for heat 15. If this does turn out to be his last ever race at Armadale, Shane closes his speedway career visits to this stadium with a comfortable win in wet conditions. Edinburgh take three points for a 49-41 victory that also temporarily slows the recent good form shown by the Sheffield Tigers. The meeting officially ends at 8.55pm though (Scott Wilson informs us) they'll segue on without delay, "We'll go straight into the second half". Almost inevitably – at 9.15pm – the rain stops as the Supporters' Club coach pulls out of the car park for the journey back to Edinburgh.

**17th June Edinburgh v. Sheffield (Premier League) 49-41**

# CHAPTER 8

# Berwick

## "There used to be one bloke in a Ford Cortina with a bike on the back and he did everything! Now they have two bikes and a mechanic, at least!"

18th June

Raffle stall

With the forecast of both light and heavy rain around the 7pm start time, Berwick's Premier League meeting against Ipswich could be in jeopardy. Around midday Berwick Bandits veteran presenter Dick Barrie strikes an optimistic note. "As I used to say when I was a promoter, 'it's on until it's off!' We'll try everything to make sure it's on because they've only had three meetings since April. No club can afford to be without cashflow – especially during their peak months of June and July. I can say for sure that they want it to go ahead!" Berwick railway station is bathed in mid afternoon sunshine while, on the other side of the river in Tweedmouth after a 20 minute walk, at the stadium it feels sultry. Out on the dark red shale of the track, there's a dry inside line. Two trucks are parked up on the fourth bend while, on the opposite side of the stadium, a man jacks up the corner of the refreshment kiosk to place bricks underneath, watched-cum-supervised by his daughter. This refreshment facility carries the words "Berwick Speedway Quality Catering". He's frequently interrupted by the chirrup of his mobile phone ringtone. There are only so many ways that you can tell people about the dry conditions and the club's intention to run ("its fine here!"). Unofficially, he tells, "We don't know what's going to happen?" Many miles away in Penicuik, Berwick fan Steven Brykajlo gives his own meteorological overview via telephone, "Berwick is on the coast so often gets the rain later or it misses completely! With the [Cardiff] GP next weekend and three rain-offs already, they'll be keen to race to avoid the fixture pileup, plus it's Ipswich, so that's a way if they have to come back." Team manager Ian Rae arrives. Pretty well all that can be done has been done so, for now, in our own ways we collectively kill time and wait for the weather. Passing trains rumble distantly, birds sing while two young boys and two young girls cycle past at high speed along the rough hewn walkway separating the grandstand from the edge of the home-straight fence, ("you can do wheelies on them bumps").

Just before four, Ian Rae fires up the blue tractor to flat grade the dark red shale of the immaculately manicured track towards even greater perfection. Anthropologist Mary Douglas famously noted that the concept of dirt is often a psychological judgement about matter out of place. Ian's attempt at greater curatorial perfection ends almost as soon as it starts with an exceptionally heavy 45-minute shower. Midway through the cloudburst – as he passes – a man in blue overalls says, "Its Saturday, isn't it?" Eventually the velocity of rainfall subsides to mere drizzle. Ian Rae looks wistfully up at the dark sky, "We didn't do any watering, let's hope it doesn't come down any heavier!" After the rain, three different pairs of riders exit the pits to inspect the track. The first pair walk its complete length and say hello as they pass. They anxiously pick at the shale with the toe of their trainers at the start line, repeat this process mid-track on bend one and at the exit of bend two. The second pair do a complete circuit but don't say hello or kick the track surface. The third only get so far as the start line for a brief tap at the track surface. Other riders wander out in dribs and drabs from the pits to head anticlockwise to around about the mid-point of bend three where they gather for a chat. Despite the rain, it's noticeable the wet look red shale still crunches underfoot when riders pass. By five o'clock, with brightness visible on the far horizon, it lashes down again. On her way to the programme booth, Davina Johnston stops. "I reckon this could be in for the night. It's not what we want. It was off a fortnight ago and, next weekend, there is none. People won't come along on a night like this, will they? All the Ipswich riders are in the pits with their stuff out though they're not supposed to be allowed in until five. We'll see, eh?"

Geoff Hayes ducks under the grandstand to shelter from the rain. We both stare morosely at the damp conditions. He's one of the men behind the campaign to get the Halifax Dukes back into speedway. "Last season I saw them ride in the Northern Junior League second-half clash at Redcar. Last night, I saw the Halifax 'Hargreaves Foundry' Dukes ride against the Edinburgh 'Kumho Tyres' Kings at Armadale – again in the Northern Junior League!" Dick Barrie works tirelessly for and regularly evangelises about the NJL. Proof that a little bit of knowledge is dangerous, page 87 of *Bouquet of Shale* has footnote on the Halifax Dukes (information I'd 'researched' on the British Speedway Forum as well as from the Redcar programme and the *Speedway Star*). Apparently, I mistakenly state "the three men behind the campaign to keep the Halifax Dukes name alive are Geoff Hayes, Peter Dyson and Dave Burnett". Geoff Hayes quickly corrects my misapprehension, "Peter Dyson has never done anything really! He said he'd sponsor us – he got some body colours made and we never heard from him again. So you saw us at Redcar and at Edinburgh last night. We only got two points in four races. We stayed in the Premier Inn at Bathgate last night. It's our Northern Tour! We go to the nightclubs." I'm surprised, "Are there nightclubs in Bathgate then?" "Yeh – three! We went to two of them. Today it were lashing down there but sunny when we stopped to jet wash the bikes. Some places don't want you to use the jet washes because the bikes are so dirty. They were after last night for sure. We stopped there last year too. Did you see that article in the *Speedway Star* by Keith McGhie? I'm annoyed as he spoke to me, because he makes it sound like Buxton started up the Halifax Dukes whereas we chose them – rather than Scunthorpe, where Rob [Godfrey] offered us to ride – so it gives us a bit of home advantage! Tamas Sike came to Buxton and he couldn't make sense of it. Tracks like Scunthorpe, Sheffield and here are easy to ride, whereas ours is tight and tricky so – when people visit – they struggle. We're what I like to call a proper team because we're mostly really all from Halifax. Well, Montana Jowett is from Darlington but – last night's team Adam Kenyan, Daniel Mitchell and Rachel Hellowell – are from Halifax! Rachel is working there and only left school in the summer last year. Paul Burnett is from Halifax, he was what we classify as our old hand. He could give them a hand and David Speight – dunno if you've heard of him – he can too and then there's Wayne Carter, team manager, who's a Halifax lad. I used to ride. I don't know about the bikes and engines and can't be bothered, whereas Wayne knows everything! We got our website up last Thursday and Hargreaves Foundry are sponsoring us. I own the name Halifax Dukes Limited. I used to be a junior out of the Shay in

1979. My middle name is Linden so I rode as Linden Hayes. It's a long story and I ain't gonna tell you! My mother took us all as kids and we got interested and I rode at that time. There wasn't many opportunities and me engine blew up at Mildenhall and I just couldn't afford to continue. In 2003 or 2004, I went to Sheffield and watched a meeting and took my sons and one of them [Grant Hayes] said 'Can I ride, dad?' and, eight months later, he were riding in the British Under-15s championship. He rode in the inaugural Under 15s and he actually rode for Weymouth at Stoke. I think it were 2004 in the last month of the season. I think he finished eighth in the Under 15s and he'd rode that season. In 2005, when Scunthorpe reopened, he went for a trial and ended up riding as reserve at Scunthorpe. In 2005 we ran a full 15-heater Halifax against Scunthorpe. Garry Stead rode for Halifax and André Compton rode for Scunthorpe. We did think we were gonna ride this season in the Midland Junior League. We joined it but, when they got down to Wolves, some bloke came up to Wayne and said 'You can't use the name Halifax Dukes!' Something to do with they weren't authorised to use the name of the Midland Junior League! Something to do with the BSPA and the bloke organising it? I can't remember. Wayne will know his name – the one who wrote all the emails – hadn't asked them at all and he admitted that he hadn't asked the BSPA or the Speedway Control Bureau. All I want to do is fulfil everything that is required by the authorities to ensure we do it right to get Halifax running!"

"Normally I spend eight months a year out of the country with work. I was in Russia a few weeks ago, China, India (for Hargreaves Foundry). Before that I worked in textiles so I wouldn't like to count all the countries I've been to but – as I say – we're the only proper team, we're all from Halifax and all are juniors! None have ridden professionally and all want to learn how to do it! Wayne Carter is the man to help them. Tonight they [Berwick Borderers] have that Arlo Bugeja. How do you say his name? I know he's an Aussie. Last season he was riding in the main meeting at Redcar where we were in the second half. People are getting to know us locally. On Thursday, I was walking down the street in Halifax at 6.30am and I heard a voice saying 'Ooh, ooh, Mr Speedway, I haven't forgotten about you!' It was the lady from Environmental Health over the noise. I had a meeting with them about three weeks ago plus the Highways Agency. We're doing it properly with Calderdale Council and they didn't object. I've not told anyone where it is – not the riders, no one, not even Wayne! Calderdale Council want to help. They say if it can't happen on that land I can't tell you about, they'll search their register for places we could have a speedway track, if it's not there. We've had a few false starts. I live at Greetland and we found somewhere. It were at the rugby club where I learnt to play and I had a noise test done – that costs £2,000 – but it were a waste of money, if I'd known what the objections really were beforehand. There were somebody called Coates (I think it were) at the Halifax Courier and he made his own news headlines which was something like 'Greetland residents against speedway' or something like that. It were only a day after I only told a very few people that that appeared. Because Coates lived on the road, he didn't want it there! I'd have never got planning permission. Grant went out and hand delivered leaflets about a meeting and one bloke said 'I didn't get one!' And we said 'Then how come you're here? Because it weren't in the Courier!' Then there were Scammonden – at one point it had the longest single span bridge in the country – and that's three miles from any house. The motorway is a mile away, so noise isn't a problem. Calderdale Council, they offered us the plan for £1,500 but, after an article appeared in the paper, the next week they withdrew it. They said that a wild orchid had been found! I was up there all the time for three weeks and there wasn't a flower! It was moorland. But then, they say, they found it on the Water Board land adjacent." With jobs to do and other people to meet and greet, sadly my conversation with Geoff ends, "Hopefully, we'll be seeing you at a speedway meeting in Halifax soon! Even if they don't get to ride tonight, it's a great experience! It's their Northern Tour and is all part of being a speedway rider. We're going home tonight although we're racing against them [Berwick Borderers] tomorrow at Buxton."

Dennis McCleary – whose job titles, let along behind the scenes work for the club, would take a separate book (or giant sized business card) to list fully – remains confident that the meeting will go ahead. "If it stops raining, the track is fine!" They open the turnstiles in heavy drizzle to signal the intention of the club, riders and officials. Early arrivals can see an immaculate, lush centre green as well as the Ipswich riders and team management out on a track walk with assorted umbrellas and anoraks. To signal he may be a long way from Australia but remains culturally committed, Taylor Poole wears flip-flops. The Witches take their track walk at a sedate pace before they huddle under the comparative shelter of the lee of the (famous 'Ducket') back-straight grandstand. After the initial rush of fans drink in the panoramic splendour of the 368 metre Shielfield Park track, they're spoilt for choice with other amenities. According to the Track Facts section on page four of the programme, there's a "covered all seated grandstand on the home straight (over 1,300 seats)", "covered terracing on the back straight", "elevated viewing on third bend (grass banking) – best view in speedway" plus, adjacent to the raffle stall, there's the Berwick Speedway Quality Catering truck aka "Fast food outlet supplying burgers, hotdogs, chips, hot and cold drinks". Though it looks closed, there's also a "Kiosk – supplying pies, snacks, hot and cold beverages (near pits fourth bend)" and to the "rear of seated grandstand" there's also a "Bar – fully licensed outlet selling a variety of beers, wines, spirits and soft drinks". In the home straight grandstand there's the "Hospitality Suite – glass fronted seated viewing area and lounge for meeting sponsors and guests to relax, complete with buffet and complimentary drinks". Completely separate to the bar, there's also the "Black & Gold Pub – situated behind the main grandstand, this licensed bar and function room is open to supporters both prior to meetings as well as after the conclusion of the racing". The "Programme Kiosk and Trackshop – fully stocked with Bandits and speedway merchandise" are both open. Jean Foster on the raffle stall looks up morosely at the dark sky and continuing rain, "This is all we need." I'm surprised by how many fans come in. Jean suspects boredom, "They've all been sitting in their cars."

At exactly 18.33, the bikes rev up in the pits. Witches fan Mark Shrimpton worries about my sanity, "You must be a glutton for speedway – watching Ipswich away!" Ipswich away at Scunthorpe is the "must see" fixture of the 2011 Premier season, so far, since barely a week goes by without this particular meeting appearing centre stage in the schedules (although, sadly, it doesn't actually run). Mark isn't pleased with the latest postponement, "Last night Scunthorpe shouldn't have been off for 10 minutes rain – that's another 100 fans alienated!" I'm not quite sure where Mark gets his figures from, let alone how he calibrates comparative alienation levels. There definitely aren't 100 Ipswich fans in Tweedmouth tonight. On his home turf, Jim Brykajlo doesn't let an element of dampness put him off. "They'll give it a go, they have to start because they've had a few political rain-offs!"

With rain heavy enough to jeopardise almost any Sky televised Elite League meeting, heat 1 gets underway. The Bandits Sebastian Aldén and Alex Edberg really attack from the outset. The racing looks wild, fast and furious in the rain. The Bandits romp away from Taylor Poole and the apparently disinclined Witches captain Kevin Doolan. Jerran Hart starts heat 2 well but soon runs out of room on the second bend in his mini battle with Jade Mudgway. Once again, it looks men-against-the-boys style speedway and the scoreline already looks uncompetitive at 10-2. The determination, speed and ferocity of the Berwick riders gainsay the apparently slippery conditions. I join Dick Barrie out on the centre green for a brief interview. From this privileged vantage point, the racing looks extra fast and the competition for those vital inches of shale much more intense. Taylor Poole wins an awarded race after Joe Haines falls and is disqualified by referee Darren Hartley. While the fluorescent clothed medical staff rush over to tend to the stricken Joe Haines, the on-duty paramedic takes a leisurely stroll across. Upon return, he explains his leisurely approach often belies good diagnoses, "The good thing is I saw Joe Haines's dad – who took one look at Joe and then went straight for the bike. I'm a paramedic. The first-aiders run, I stroll over calmly! It gives me more time to make a diagnosis. I've been doing it six years in the North East. I go to lots of events. This is my second year at speedway – I'm still learning the rules! It's great but it's my first time in the rain." Lasse Bjerre races in exhilarating fashion intensely pressured throughout by fast-riding Bandits captain Lee Complin poised within an inch of his rival's wheel as he vies and searches for the vital glimmer of an opportunity to pass. Sadly Complin's pursuit ends when he runs out of tear-offs.

Experienced start marshal Tom Anderson has no worries about the Shielfield Park track, "It's better than lots of places – it'll take three or four races before they get a line on it properly!" Heat 5 requires a rerun and sees the Josef Franc/Kozza Smith partnership race to the Bandits third maximum heat advantage. Kevin Doolan picks up a consolation point but still looks a reluctant participant. Paramedic Paul watches intently, "They have to be brave or warped! I think crazy! I was naive when I first came. I expected it to be a small crowd but they can be four times what the football does. It's the biggest sport in Berwick for sure, I think it's great! I bring the kids. They love it! They've been mascots and out on the centre green. The lassie is 5 and my boy is 12 and they're sitting in the crowd." Speedway is certainly a child-friendly sport. If speedway faces trouble ahead, then long time start marshal Tom Anderson has no doubt about the culprit. "Everything the Speedway Control Bureau has done has ruined the sport and added more expense! There used to be one bloke in a Ford Cortina with a bike on the back and he did everything! Now they have two bikes and a mechanic, at least! There's silencers now and the tyres – well just don't get me started."

Out on a tactical ride in heat 7, Lasse Bjerre starts like an express train, albeit tailed by a determined Lee Complin as the metaphorical coal wagon. Though home track knowledge is a significant advantage tonight, mechanical gremlins enable Complin's second lap burst into the lead past Bjerre. By the time he reaches the last lap third bend, Lasse Bjerre merely coasts so can't quite coax his machine home over the final yards ahead of Mitch Davey. Paramedic Paul sympathises, "These conditions and the dirt could have thrown up something into his bike." Though not quite an expert yet, paramedic Paul knows what he sees, "We had a tall gangly one last year and he looked like he was always going to come off!" He also enjoys knowing some of the riders as a result of his duties, "Craig Branney has a training school at Northside at Workington now. He was a nice lad but unlucky. Every time he came back I saw him [on the track injured]." Heat 8 is worth the proverbial admission money alone. Taylor Poole and Jade Mudgway race side by side throughout laps two and three. No quarter is asked or given. Bizarrely, after the race Mudgway – who I thought rode the black and white helmeted Poole extremely hard – gestures his annoyance. Start marshal Tom watches approvingly, "Poole showed great control in the bends to keep a smooth line! It would have been easy to have wobbled but he didn't, he's gonna be a good rider! His dad was good too." Tom Anderson is also quick to praise Lasse Bjerre, "His brother is good and you can see he has a good pedigree."

Whoever Ipswich sent out for Morten Risager's rider replacement ride, easily has the measure of Berwick's Alden/Edberg power partnership. Dick Barrie identifies the mystery rider without pause, "It looks like Doolan to me." Heat 11 sees Kevin Doolan's gating gloves work in successive races. Effectively no longer really a contest at 44-22, the next race sees Jade Mudgway (sometimes described as a "rugby player on a speedway bike") battle for second place with Chris Mills. Kozza Smith wins and his post race celebrations see him park up by the start/finish line and leap muddy kevlared into the home-straight grandstand (potentially guaranteeing some delighted young fans speedway memories for a lifetime).

Though the final three races of the night are competitive over the initial laps, the spark has left the Ipswich Witches. Margaret from Edinburgh sums things up, "It's gate and go!" Sebastian

Aldén wins heat 15 to record a paid maximum. Ignoring he previously rode at Swindon, Dick Barrie says, "He's a real number 1 – I'd like to publicly thank our manager with the dicky ticker, Ian Rae who sourced and discovered Seb Aldén!" Despite the rain, it's been a lusty and often hard-fought battle. Jim Brykajlo is a hard-to-surprise long-time Berwick fan. "Some fans leave and some fans linger, the crowd was crap tonight because of the weather! I got in for a tenner because I had a 1970 Ipswich programme from the KO cup final at the end of September. We lost the first leg there 47-31 – John Louis rode and we won the second leg 43-35. I got in free at Plymouth earlier in the season. Anyone who had the Plymouth programme from August 22nd 1969 got in free. Only two of us got in because we had it! I have every Berwick programme since 1968." It took quite a search according to Jim's son Steven, "Once it was on the internet, he was in the attic for about three days trying to find it!" Though many fans rush off, Dick Barrie alerts us that "half price burgers for the rest of the night" are available at the Berwick Speedway Quality Catering kiosk. One of those to join the queue is a slightly bedraggled looking Taylor Poole dressed in traditional Australian speedway rider style: flip-flops(!), shorts, T-shirt and, as a concession to the rain, his baseball cap from underneath which damp frizzy almost Pre-Raphaelite hair surges like a burst sofa. Departing Edinburgh fan Mary Kelly identifies referee Darren Hartley's start gate technique, "This ref is one, two, up with the tapes!" She despairs that BSPA chairman Alex Harkess didn't decisively intervene during the disputatious close season, "They should have knocked heads together during the winter!"

With Berwick riders out on the centre green, Dick Barrie reminds us of the prize ("bottle of the finest bubbly") given to the most entertaining rider of the night. The steward close to me remarks, "If Seb Aldén doesn't get it I don't know what they're thinking!" In fact, for the first time ever, "It's a three-way tie between Jade, Seb and Mitch". It's news that slightly nonplusses the steward, "Mitch Davey did well I agree. Seb was excellent I also agree but Jade why? Everyone says Mudgway is so good but Sike didn't have a chance at number 6. Someone tried to explain it to our glorious leader, but he didn't understand that your best reserve should be at number 6. If you look at the programme: they've got reserve, not, not, reserve. Sike kept getting taken out of the fourth ride against the reserve for Mudgway who, obviously, did okay against other reserves!" Ever professional Dick reminds us, "Don't go away, the night is young and the burgers are cheap!" Though the weather cuts short the post-meeting centre green interviews, Dick grabs Jade Mudgway to discuss his 15-year-old brother. Brady's over in Europe to compete in the F.I.M. 250cc Youth Gold Trophy Championships in Scandinavia, "Can Brady gate better than his big brother?" Jade verbally shrugs, "He can't do worse!"

Though tonked 62-28, Dick Barrie congratulates Ipswich, "I would like to publicly thank Chris Louis and Pete Simmons for being exceptionally co-operative visitors!" Bandits promoter John Anderson stops by my table, "We had two years without any rain-offs and now we've had some!"
[Jeff] "You did well to get it on."
[John] "We didn't water it today, that's why. How did you do?"
[Jeff] "I sold one."
[John] "That's shit."

The meeting officially closes at 9 o'clock. Jim Brykajlo remains in upbeat mood. He believes the staging of tonight's meeting sends out a message to stay-away or wavering fans. "Berwick proved they can get meetings on when it's wet! And have good racing! The racing didnae justify the scoreline." As he exits, one member of the small but perfectly formed group of Ipswich fans at the meeting says, "We were the only problem out there – the track was fine! We have a psychological problem and we're getting better at it!" Twenty-four year old Gavin Ritchie came along with the Brykajlo family tonight, Jim tells me, "Out of about 50 people I've brought to speedway, Gavin probably is the only one who's stayed." Luck plays some part in whether people return, according to Steven, "He got lucky to come to a good match. I think his first match was away at Redcar versus Berwick in 2009." After three or so years on and off, Gavin knows what appeals to him about Berwick, "There's a good atmosphere and it's something different! It's good banter but not always good racing and the Berwick burgers are good." Hooked on all things Berwick Speedway, Jim also enjoys the refreshments. "They're my favourite burgers at any speedway track. They come from John Anderson's shop – they kicked out the guy who'd been here for 30 years. If you think about it, if you can sell over 100 burgers every week of the season – it's good money. He needs the money he's making flying the two Swedes in each week to Edinburgh. Swedes are always the most expensive in the league because they ride on a Tuesday night."

Jim's love of speedway usually prompts further reflections, "I think Mike Bennett is an excellent presenter. Then I like his videos and DVDs too. I'm one of the biggest purchasers he told me once!" It's a moot point how many fans the almost radioactive egotism of Mike Bennett attracts to speedway. Like gateway drugs everywhere, you're more likely to get hooked through the influence of family or friends. Jim warms to this theme, "How do people go? Matthew Platten went with his dad. How did Steven go? Because I took him and my brother George I took. How did Dougie Copland go? Because he's my cousin – Dougie's mother is my mother's sister, my aunt – I took him to Berrington Lough when he was seven or eight years old. He was a Berwick fan but he doesn't like to admit it now." Steven reveals Jim has his own dark secrets, "Bet you'd never admit you were a Monarchs fan in the 60s!" Unabashed Jim continues, "It's a generation thing." [1]

**18th June Berwick v. Ipswich (Premier League) 62-28**



[1]The Borderers beat the Dukes 15-8 in a second-half NJL meeting – completed before 9pm - notable for exciting races and the total absence of Arlo Bugeja.

# CHAPTER 9

# Glasgow

## "In those days, we didn't talk to the riders"

19th June

Damp red shale

According to gossip at the pits gate, my Father's Day trip to Ashfield Stadium is threatened by forecast rain. From the puddles that fill the potholes of the car park and rough hewn track vehicles use to access the pits, this part of Glasgow saw earlier heavy rain. With new teams from above (Ipswich) and below (Plymouth), the 2011 Premier League is ultra competitive. That's certainly the view of Ipswich co-promoter Chris Louis, "Our fans have taken to the Premier League very well and from our point of view it's more straightforward and very enjoyable. We were struggling financially in the Elite League and will remain in the Premier for the foreseeable future. It's a good product and at the moment there are five or six teams who could win it this year given good fortune."

So far this season, Glasgow look strong. Consequently, fans, riders and management alike are (guardedly) optimistic. Though teething problems saw initial home fixtures against Newport and Workington postponed, it's easy to see winter work made the Ashfield track bigger, wider, far more banked on the third/fourth turns. It's also been restored to the more traditional oval shape you'd expect from a speedway track. Riders and fans alike are positive about the changes since there are improved racing lines and, it's said by regulars, greater spectacle. It's clear that the track boundary walls moved, plus the track topology and surface altered. There's still substantial work to be done in and around the stadium. The programme notes club landlords Ashfield AFC are about start work "enlarging the football pitch which was part of the agreement when we made our track alteration". The panoramic view from the white bricked bar area that overlooks bends three and four appears severely diminished. Behind comparatively see-through red gauze and temporary protective fencing, there are numerous closed shutters and only two open windows. In fact, though it looks the builders abandoned their work half done, recent strong winds caused this storm damage to the bar area. It ripped off the family room roof and cut the stadium landline.

Over the stadium speaker system, club presenter Michael Max strikes an optimistic tone, "A warm welcome to a slightly soggy Ashfield. The pits area is open but, if you are thinking of coming over, just to warn you that conditions are a little gooey! It is our intention to start at 4 o'clock. The ref [Peter Clarke] is happy that there may be a need for a little remedial work!" Berwick Bandits fan Jim Brykajlo has followed speedway long enough to know meetings can race in the rain or get called off. "If that had been Glasgow or Edinburgh last night [at Berwick], they would have called it off earlier! But, since it was Ipswich, they went ahead. I thought the racing was good and better than the scores suggests." Life-long Glasgow Tigers fan Ian Maclean is a reluctant meteorologist, "Michael Max can see bright sky when there isn't any! Today there is. The forecast was okay but it is wet!"

It's not only the track that's allegedly held Glasgow back over recent seasons but also a propensity to attract injuries. This season the club continues to find riders out of action in its metaphorical treatment room: Josh Grajczonek (broken femur), Nick Morris (broken collar bone), Christian Henry (broken shoulder) and Joe Screen suffers from tendonitis of the elbow. Nonetheless, the Tigers continue to perform well home and away. Ian Maclean's son saw last night's Glasgow visit Beaumont Park in Leicester. "Alan says James Grieves was magnificent last night. He's posted on the Growls saying that. I saw James and said, 'Do you ever look at the internet about yourself?' and he said he didn't! I said, 'You should go onto the Growls because rather than nasty things being said about you Alan has put something nice!' When they were at Edinburgh there were some there who didn't like him and were saying things about James. One thing's for sure, they wouldn't say it to his face!"

With 30 minutes to go before the scheduled start time, Tigers diehard Alison Chalmers tells me, "The photographer says that the riders aren't changed and they're not getting their bikes out! We were going to have a team photo at 3.15 – with the young members from the Growlers Club – but conditions in the pits aren't good, never mind the riders aren't changed! I'm not here for five weeks now – I want it to be on! What a win last night! The supporters' coach was cancelled at the last moment because they couldn't get a driver but lots of fans still went." With 15 minutes to go, it rains heavily. Allegedly not a big fan of wet shale, Ipswich Witches captain Kevin Doolan is already in his kevlars. Kevin carefully inspects the start-line area in the careful manner of a crime scene investigator and then wanders to the new look wider bend two to join a gaggle of riders in a conflab with the referee. Even Michael Max struggles to find an imaginary silver lining, "There's been a lot of rain over the last five minutes or so, I'm afraid the meeting has been called off!"

The track looks much improved but, covered in puddles, isn't raceable. No sooner is the meeting abandoned than the rain stops and the sky brightens. Purple Rain blares out over the stadium tannoy to serenade departing fans. Kevin Doolan picks his way back across the sodden track towards the changing rooms. Ian Maclean calls out, "I was looking forward to seeing you ride today." Affable and accessible like speedway riders everywhere, Kevin smiles, "I was looking forward to having a skid but what can you do?" Ian's appreciative, "I've always admired you." Kevin grins again, "Thank you!"

By 4 o'clock, it's a lovely afternoon. Minimal sunshine transforms the track. Donald Forbes studies the red shale surface wistfully, "Look the track is perfect now! I said why don't they wait until 4.30? If they had some shale – which they don't – they could put it down on any tricky bits. The riders are wimps nowadays! The only problem is turn two because they didn't put in drains there when they did the work, so all the water runs down the hill and gathers. If you're looking for Alan [Dick] he's in the pits paying the riders!" Ian Maclean tells me that Donald Forbes "rode second halves – I know he's been on bikes in the late 40s at White City." Talk of yesteryear soon takes Ian back down memory lane to the Ashfield Stadium of his youth. "At 15, I was the only boy in my class not to have been to speedway! I thought it was motorbike racing but thought 'what's the point of that?' I'd heard them talking about it for years. We went – my friend Fred and I – to the first meeting of the 1952 season versus Coventry at Ashfield. It was totally different then without most of the buildings they have now. It was a giant bowl and we queued for ages and, as we got in, the riders entered the first corner all together. It was the first sight I ever saw at speedway. I was amazed and hooked from the first moment! I can remember the score too. It was 42-42 or 44-44. There were 14 heats. We were 12 points down with three races to go and Fred said 'That's it!' But we won 5-1, 5-1 and even I got carried along, by the excitement of the crowd, with the last 5-1! Ken Le Breton was dead by then – it was 1950 – I remember two girls crying in chemistry class. I couldn't understand it because they didn't know him but I understand it now! In those days, we didn't talk to the riders. Even in the 60s we didn't and we didn't go the pits!"

**19th June Glasgow v. Ipswich (Premier League) – postponed**

CHAPTER 10

# Birmingham

## "'Make Me A Fan' scheme. It's called promoting!"

23rd June

New technology

Birmingham Speedway's Elite League arrival is a widely seen testament to the drive and ambition of their promotional team of Tony Mole, Graham Drury and Mick Bratley. The Brummies early season success is built upon the hard work of staff, volunteers and the riders alike but also relies upon the behind-the-scenes work and expertise of Graham and Denise Drury. Though at the club since the get go, during the close season it was announced that they would reluctantly leave the club for cost reasons. No sooner did this hit the news than a kindly sponsor made an offer that enabled an immediate reverse ferret and their reinstatement.

Three hours before the first race, Perry Barr stadium bathes in warm sunshine cooled by a gusty breeze. The sky belies the mid afternoon forecast of heavy thundery showers. Inside the Speedway Office close to the almost empty pits area, Graham, Denise and referee Jim McGregor chat between endless phone calls. Graham says, "I mentioned in the local press last week that you'd be here". Outside the office, there's a half unloaded van (without occupants) liveried in Edinburgh colours emblazoned with messages about the club on its bodywork (Send your message: "Every FRIDAY 7.30pm Armadale Stadium") plus a generic broadsiding Monarchs rider photo. Out on the track, staff are half way through their powerwash of the bends one and two airfence adverts.

In the main home-straight grandstand building, the trackshop has a different table layout. After many boxes of merchandise are unloaded, Nick Barber asks, "Have you seen the new masterpiece? Shut your eyes and hold your hands out." It's a giant badge with rough diameter midway between a squash and tennis ball (made specially for the Cardiff SGP). "If you put it on your lapel you could potentially fall over!" Given the older demographic speedway enjoys, it's conceivable fans still own and wear clothing with lapels. Sadly nowadays, such is the casual fashion malaise of modern life (plus the rise of the branded speedway anorak) – even at speedway – smart jackets are no

longer worn frequently. Nick thinks his badges should be worn on a lapel. Four pins hold them into position, whatever garment you choose to wear. Feedback isn't totally positive, "Robin – the biggest badge collector in the UK – said, 'I don't like it!' He don't like heavy badges." Different coloured badges features different SGP speedway "superstars". Nick's sister, Bev studies a badge critically, "The faces haven't come out so well. We did one for Oxford and there was only seven riders – so it was much more detailed!"
[Nick] "And they didn't have their caps on."
[Bev] "Its three inches round because I measured it earlier."
[Nick] "It's 70 centimetres in diameter."
[Bev] "Eh?"
[Nick] "Seven centimetres – that would be more like a manhole cover than a badge."

Earlier in the week, either administrative oversight or serial incompetence led BSI/IMG – organisers of the Speedway Grand Prix Series – to belatedly announce the epic fail that Artem Laguta hadn't the visa he requires to actually enter the UK to ride at the Cardiff SPG! More suspicious minds believe that talk of "visa problems" is a smoke screen to obscure deeper-seated chicanery and other unspecified (possibly financial) shortcomings. Luckily, Artem will make it to Cardiff in badge form, "Yeh, but don't ask me which one he is!"
[Jeff] "I see Barry Briggs is coming to do your Fayre at Cardiff."
[Nick] "Is he?"
[Jeff] "It says he is in the Star."
[Nick] "He's been ringing me all week. I've called him back four times but I haven't got him."
[Bev] "Where shall I put the pins then?"
[Nick] "Not there! You can't see the riders then. It's spatial awareness, Beverley."
[Bev] "What's the point of there? You can't see the riders!"

The Speedway Fayre at Cardiff Rugby Club is the premium sales opportunity of the speedway merchandise year. High expectations abound, "Gary Patchett rang me Monday and asked if I could do a hundred Leigh Adams T-shirts. I told him I could sell them but not make them in time for Cardiff! Dunno what he was thinking?" Outside, with well-deserved cigarette in hand, Bev surveys the low clouds and dark horizon. "Unfortunately the clouds are coming this way but, as Nick says, it'll be a sharp shower and then there's bright sky behind it." Tony from Clean Cut DVDs embraces the spirit of the summer but also must feel flush since – at £2 for a small "99" – he's bought one of the most expensive ice creams in British speedway! "I'm having an ice cream before it starts chucking it down! It's meant to be really nice at 10 o'clock tonight." Briefly donning his metaphorical Eastbourne promoter's hat, Nick Barber wonders, "Cameron Woodward got through the GP qualifiers and then got one point at Poole – what's that about? Get your priorities right, Cam!"

Like their fans, Brummies co-promoter Mick Bratley feels a strong sense of déjà vu, "It's our third meeting on the trot against Peterborough! Two aways and, now, at home. People ask me 'Have you got divided loyalties?' No I fuckin' haven't! I wanna win and I wanna win big. I'll chat later about my 'Make Me A Fan' scheme."
"What?"
"'Make Me A Fan' scheme. It's called promoting!"
"Not everyone has heard of that."
"That's good, [makes a circle gesture with his hands] we only want it to be round here. What's your new book?"

"*Bouquet of Shale*."
"Not many will be able to say bouquet!" [1]

By the double doors that used to be the main entrance to the stadium, Peterborough Panthers owner Rick Frost and wife Julie Mahoney loiter with their daughter. Mick Bratley nods over in their general direction, "They came to the wrong entrance. They should go to the pit gate. There's something very satisfying about making Rick Frost sign in!" Denise Drury helpfully chases after the Frosts to help smooth along their first ever official visit to Perry Barr Stadium, "Do you want a programme? We left some for Trevor [Swales] over at the pits." Rick Frost lingers and chats amiably to fans. "Who's to say it's over? There could be more legal action. The problem is that some people just wanna keep their hands on the money and don't care about the sport! The management of the sport needs a spring clean. "
[Jeff] "There was so little news and it was often contradictory. The BSF became readable because it was so partisan."
[Julie]"I felt sorry for the fans because we knew what was going on but, obviously, they didn't really and we couldn't say anything."
[Rick] "The trouble is – unless things change – the sport will be destroyed! Some people don't care so long as they're alright. Don't expect too much of us tonight, we have a patched up team."
[Jeff] "The good news is that you're still running."

While he chats, fundraising fans hand Rick a three-page stapled document about the membership benefits of Birmingham Speedway Supporters' Club along with an application form, pre-printed standing order mandate and a full page of rules and regulations. [2]

Because of holidays and staff shortages, Clean Cut's supremo Pete Ballinger operates their sales outlet "I'm not used to doing this but Tony's brother Andy has gone to the Isle of Wight. When I was on the way here there was a huge black cloud over Villa Park."

[1] "Bratley's Bulletin With Mick Bratley" carries news of more promotional initiatives at the club. "I'm pleased to announce to you that plans are now in progress to establish a Birmingham Speedway Hall of Fame here at the Perry Barr Stadium. I have recently held discussions with the stadium management and they have been very accommodating in allowing us to do this, and in fact very supportive. The Hall of Fame will be located on the first floor foyer of the grandstand. The plan is to induct a "team" of eight riders initially and this will be undertaken in liaison with the Supporters' Club, Tony and Graham. Thereafter one or two riders will be inducted per season and this can be done via a vote system amongst supporters. I like to think one or two of you will put forward a case and lobby/canvass for who you think should be in our Hall of Fame. More details will be announced soon, and rest assured we will be commissioning something that is very professional and something that all fans will be proud of, something that is a fitting tribute to the great riders of our past. If you would like to contribute financially to the Hall of Fame and have your name on a plaque of recognition as a lasting tribute then please contact me."

[2] There are three categories of membership:
**JOIN TODAY – ONLY £5.00 ADULT MEMBERSHIP/£3.00 CONCESSIONS**
Exclusive Social Nights with management and riders
Discounts on 'away travel'
Summer and Christmas parties with AMAZING PRIZES donated by the riders, management and trackshops
Discounts at Clean Cut Productions for DVD's
Competitions – get your painting in the BSSC Calendar 2012
100 Club – £2.00 per month per share – also open to non-members
**SKIDZ Club – Only £3.00 membership for kids aged 5-15 years (Under 5's FREE)**
Birthday card signed by the team
Photo opportunity with your favourite rider
Weekly prize draw at 'Home' meetings with photo appearing in the programme
Competitions – get your painting in the BSSC Calendar 2012
**COMING SOON – exclusive SKIDZ Club badge – free year bar on membership renewal THESE BADGES ARE NOT AVAILABLE FOR SALE**
In addition to these benefits there are a number of attractive forthcoming events including:
GP Night – Saturday 23rd June at Aston Manor Cricket Club
Summer Party Friday 15th July with disco, games, BBQ and amazing RAFFLE
Overnight trip to Poole Wednesday 17th August
End of season review with Graham Drury
Christmas party Saturday 10th December with disco, games, buffet, Secret Santa & another AMAZING RAFFLE – with Father Christmas in attendance
Cut off date for Calendar competition Thursday 18th August 2011
Beer and Skittles Night – date to be confirmed

[Jeff] "Could be there for a very long time."
[Pete] "It's because Take That are playing there tonight."
[Jeff] "They've already had their tour highlight – their first gig!"
[Pete] "What, the Stadium of Shite?"

An extensive check of the powerful stadium speaker system interrupts conversation throughout the grandstand. We collectively receive what sounds like our own deep breathing phone call. Just before the breather approaches anything like a happy finish, they switch from frisky to cleaning mode, if judged by the impressive blow-the-dust-off-the-microphone-and-surrounding-area repertoire of huffs and puffs they conduct. Finally they close by barking out a random number in true early punk fashion, "2! 2! 2!" Sound check over Pete Ballinger resumes, "The big story of June is rain-offs! I've had eight or nine rain offs since May 30th. I'm not going to Somerset tomorrow because it's a Super7 event – we might go to Cardiff, I don't know yet – because we've got to go to hospital with Daniel, my six year old son, to have his arm checked. Two months ago he was whizzing around on his scooter pretending to be Nicki Pedersen when he fell off. It's the first time ever I've canned the filming. It was a close meeting but shit happens. We took him to A&E and found he'd broken his humerus. We've been going back every two weeks to check it's setting properly." Though he spends his life behind a camera filming others, Pete's a reluctant subject. "You can't have my image! I need to lose about two stone. It's IMPOSSIBLE at speedway. If you work outside and get cold – even if you're not hungry – you have to have a burger or something, whether you want it or not!"

In the early evening sunshine with a friend, Nicki Pedersen's wife Anne Mette Hansen drink in the atmosphere. Brummies presenter Chris Simpson is well *au fait* with new technology. In fact, he's at the cutting edge of the speedway presentational world with a bespoke racecard preloaded onto his iPad. He happily demonstrates its capabilities to me with a few fictional race results. When you enter heat scores the points automatically tally in the racecard, while any rider changes alter with impressive speed and efficiency. "It's really good except if it rains!" I wonder, "Or is really sunny?" Chris bats off my meteorological objection, "That's fine! This is on full brightness – so would be alright!"

A Brummies fan wearing so much red and yellow branded clothing, that he surely must be one of the trackshop's best customers, he says, "We were worried when we came up to the Elite League that the Number 1s would be miles out in front but, in fact, the racing has been good so we can't complain!" In fact, there have been many complaints about Brummies number 1 Krzysztof Kasprzak on the internet forums but also from club promoter Graham Drury. The gifted bottle-blond Polish rider blows hot and cold but mostly cold. After making a forecast ("My early doors prediction is 52-38") Pete Ballinger turns to more serious matters. "I read your blogs on the GPs and I have to say it rubs me up the wrong way! I agree with what you say on Sky – they over-egg the pudding – but the negativity about the GPs is out of order! Think back, what about Pocking and no airbags and Ulamek going through the fence and half way up the terraces? Sure BSI are in it to make money, there's nothing wrong with that! I'd like to see us get to Russia, Bulgaria, lots of other places in the world. What BSI have done has revolutionise the thing and should be praised and admired. I do and so do many others. What would the BSPA have done with it? What do you think BSI should do that they aren't doing? Not much I'd suggest." Ignoring the sheer staleness of the competition, how BSI gerrymander rider selection, fiddle with the format and now have too many GPs in inadequate locations, I reply, "They're farming what they've got and preaching to the converted and not getting the sport recognised and appreciated by any new audiences. We're supposed to be a big name sport yet there's never any coverage in the recognised media. How long have IMG had it now? They're supposed to be the best sports promotion company in the world! They certainly are with other sports."

"Maybe the people at Sky say you've got a rubbish product! That could be a reason they haven't? Let's hold our hands up and say speedway can be rubbish! If Nicki Pedersen wasn't here tonight, most wouldn't bother to come. I couldn't care what you say about Sky. They deserve everything they get for saying the racing is brilliant when we can all see it's hopeless! But BSI have done marvels for the GPs and I don't think anything you say is really that fair!"

Nicki Pedersen is still a big attraction at every meeting he races in. Though he's out in heat 1, his teammate Linus Sundstrom gates leaving Nicki to battle with new Brummies arrival, Ulrich Ostergaard. Krzysztof Kasprzak is comprehensively last and stood next to me Mick Bratley exclaims as the riders exit the second bend, "Oh, for fuck's sake!" It takes Nicki over a lap to establish dominance over Ulrich, something unthinkable only a couple of years ago!

Mick Bratley is out and about inside the stadium. He holds a bucket (of chewy lollies) decorated with the image of Lord Kitchener emblazoned with the words "Make Me A Fan". His Buddies Plan already enjoys good success according to his programme notes, "We have now gone through the 100 barrier for the 2011 season and are well on the way to the start of season target of introducing 200 new and lapsed fans to the 2011 Brummies." Before Mick shoots off to drum up yet more interest from the Birmingham faithful my question ("How's this for a crowd?") sees him first look round with pantomime ostentatiousness, "I'd say about average."

Giving us the race results, Brummies announcer and programme editor David Rowe rolls the 20-year old Swede's name around his mouth like an exotic taste ("Lea – nuss") before we learn his 57 seconds is "the fastest time so far this season at Perry Barr." If Chris Simpson's iPad aids his presentational work, then Pete Ballinger's wife loves to look on Twitter. "My wife spends all her time looking at the riders on Twitter. I don't have time because it's edit, edit, edit. Every morning she tells me all the latest news. For example, Lindgren got hit in the knackers last night or Holder wanting to know where his £5k for the EL Pairs is? When you think they only had about 800 there, it's no surprise the BSPA cut the prize money!"

Birmingham immediately hit back with their own 5-1 through Wolbert and Vissing to level the scores. Daniel King wins heat 3 but it's a race much more notable for Ben Barker's lap 1 substantial backstraight wheelie that ensures he never regains the ground he loses to Troy Batchelor. Before Heat 4 gets underway, Chris Simpson playfully lambasts absentee fans, "It's absolutely gorgeous out there tonight, those of you with little faith!" Panthers heat leader Krzysztof Buczkowski battles with in-form Brummies reserve Claus Vissing until the exit of the second bend and then appears to throttle off to settle for third. Though possibly he has the excuse that it's his first ever visit to Perry Barr, it's an easily conceded maximum heat advantage. From inside one of the hot toilet cubicles a child's voice says, "Are you using the toilet, daddy?" "Nah, I'm standing here looking decorative." "Are you dancing?" "I'd dance if Kasprzak wins a race. [two minute warning sounds] Come on you, two minutes!"

Sadly the gentleman and his child disappear into the crowd before Krzysztof Kasprzak wins heat 6, so I miss his dance. One joy of speedway meetings at modern greyhound stadiums is that, often, you can immediately watch replays of the race on their television monitors. Pete Ballinger casts his professional eye over the race action before he exclaims, "I think that's the worst spelling of UlRIK OSTEGUARD I've seen! It's wrong in every department!" The initial attempt to run heat 8 sees Claus Vissing rifle away from the tapes. It's a sight that delights Pete Ballinger, "Bloody hell! Vissing is away like a rocket. No! He hasn't stopped it! [the ref has] I thought it was just a good drop." The power and the glory that is the magnificence of the BSI/IMG run SGP Series continues to play on Pete's mind. "A case in point about BSI, I was making Daniel's appointment for tomorrow and I mentioned Cardiff and the receptionist said, 'What, for the Speedway GP?' Obviously she lives in Coventry and knows all about Coventry but that shows how much they've done! Who would have said after the first Cardiff it would still be there a decade later?" Second time out Linus Sundstrom enjoys a Heat 8 "tapes to flag" victory. Before he gives out the race result, announcer David Rowe issues urgent instructions, "Just a message for the ground staff. If they could attend to the advertising hoarding – that's blown about half way across the centre green – that would be much appreciated!"

With regular travel to various speedway clubs on a weekly basis, Pete is familiar with the good, the bad and the ugly of sport including its presenters. "Mike Bennett is harmless but there are far worse, like Peter York. He just gets up and speaks into the microphone. He doesn't watch the racing and he gets things completely wrong! I don't reckon he even reads the comic. He probably doesn't even know it's Cardiff this week!" Heat 9 sees Brummies Barker and King combine for a maximum heat advantage to further stretch their lead to 33-21. Panthers team manager, Trevor Swales sends out mixed messages in heat 10. He gives Troy Batchelor black and white but also sends out Richard Lawson for his second rider replacement of the night in place of the absent Kacper Gomolski. "Lawson has zero from four rides, yet they bring him out for his second r/r ride."
[Pete] "How many has Alex Davies got?"
"One."
"Well there you are."
"To be fair, he did look good in his last race."
"He might beat KK! Very strange his loss of form!"

To celebrate Troy Batchelor's win and KK's third (ahead of Lawson), Pete Ballinger buys a burger and chips from the nearby kiosk. Since he's been indoors all night I ask, "Are you cold?" The food is soon gone, "It filled a hole. I think I might have a lager now to sweat it out!" For the first two laps of heat 11, Daniel Nermark shows temerity to lead Nicki Pedersen until a couple of hard corners quickly persuade him that discretion might be the better part of valour. Nermark slackens off so much that Linus Sundstrom snatches second place on the line. Successive heat advantages (one tactically inspired) suddenly narrow the scores to 37-32. Pete Ballinger shrugs, "There goes my prediction, there's only 5 points in it!" Possibly suffering a temporary glitch on his iPad, Chris Simpson garbles the heat 12 race line-up. From the referee's box announcer David Rowe politely requests, "Can we do that one again?" Chris apologises, "My finger must have slipped." After Barker wins heat 12, Chris Simpson tries to drum up enthusiasm for the British Grand Prix, "Don't forget you can see Ben Barker at Cardiff [pause] he's the second reserve!" Any diehard speedway fan will already have their Cardiff SGP ticket. News that Barker is to spend the day bored and kicking his heels in the pits deep inside the bowels of the stadium is highly unlikely to prompt additional ticket sales. At 20.32, they take a 15 minute interval. My observation,"1 hour and 2 minutes for 12 heats" fails to light Pete's fire, "It seems longer than that!" Nick Barber experiences slow sales, so far, tonight but my explanation ("It's before Cardiff – people don't have the money") fails to convince, "I don't think I've spoken to anyone tonight who's going to Cardiff!"

With the riders' toing and froing at the tapes prior to heat 13, Nick Barber offers quietly spoken advice, "Don't ride like you did in the last race, Daniel!" Sight of the fast-gating, fast-charging, effortlessly stylish but aggressive Nicki Pedersen has been in disappointingly short supply over the past few seasons. Nowadays many GPs find Nicki in a fight with his sense of entitlement and apparently malfunctioning equipment. Nicki spills onto the deck or into the airfence on a much more regular basis than even his harum-scarum close-contact racing style dictates. His personal man versus machine battle has carried on into recent EL televised fixtures and, it wasn't till his race win last time out, that flashes of the old Nicki we so relish returns. He certainly flies from the tapes, so much so I remark, "Nicki finally looks quick in this race!" Phlegmatic as ever, Nick shrugs, "If he doesn't go through the card at a meeting like this, there's something wrong! He hasn't mind." Out on the centre green Chris Simpson surfs the worldwide web on his iPad, "Tonight – surprisingly –

we've only had one check-in on the Birmingham Facebook page! So a big shout out to Elizabeth Abbott!" Daniel King pressures Troy Batchelor throughout the last lap of the penultimate race but can't find a way past. Chris Simpson peers at his sun-dappled iPad and wonders aloud, "Was that Troy Batchelor in yellow or was it Lawson?"

Prior to the rider nominations for heat 15, David Rowe passes on a reminder. "Message for the track staff. Can you bring your red flags to Keith Kershaw at the start line so that they can be transported to Cardiff?" [3] When the tapes rise, according to Nick, referee Jim McGregor is lenient, "Ragged start, ho ho he's let that one go!" It's a decision that appears to disadvantage Nicki Pedersen since he exits bend two in third. However, Nicki then powers down the back straight to blast round close to the airfence and, as he transitions from bend three to four, he executes an incredibly sharp high velocity cutback and surprise mid-track shimmy between Barker and King. He gains the lead but also arrives at unfeasible speed into the first bend of the second lap. Riders without Nicki's skill would plunge head first into the inflatable safety furniture but, with theatrical acrobatic skill, Nicki hangs onto his bike to scythe away up the back straight for victory. It's an exhilarating racing manoeuvre that's worth the proverbial admission money alone. On the stadium monitor replays, it looks even more spectacular. Not only do Barker and King drift wide to block his anticipated outside run but they also co-ordinate their on track positions to such extent that only a gifted racer like Nicki would find said proverbial illusory gap at such high speed. Though over in a moment, the audacity and verve of Nicki's overtaking manoeuvre quickens the pulses. Pete Ballinger believes referee Jim McGregor delivers some justice, "Batchelor tried it and messed himself up!"

Sales are slow but, at least, I get to enjoy sight of a fan in a coloured T-shirt emblazoned with the words "Premier League Fours Champions 2010". Despite a win in his paid 7 points tally, some sections of the Brummies faithful still disparage Krzysztof Kasprzak. An old age pensioner lady querulously rages, "If they don't shoot him, why have a gun?" Pete Ballinger isn't impressed, "It shows how bad things are when I say paid 7 is an improvement. You should have seen him at Peterborough last week. Two points! He was woeful. He didn't beat anyone, only Ulrich. When you think that last year he broke the track record for Coventry!" Pete is such a fan of BSI's SGP series that even though he's yet to see it, the 2011 Cardiff Grand Prix programme delights. "It's £10 a programme at Cardiff – but you get a free radio tuned into a local frequency to hear the world feed." Quite how you'll hear them above the din of air horns is a moot point. While, perhaps, the expert analysis of Jan St chmann and sidekick Darren Fletcher is a benefit, it's not so long ago that a (comparatively) less rapacious BSI included the programme with the cost of admission!

Birmingham club physiotherapist (and co-team manager of Belle Vue Colts) Steve Williams contextualises my lack of sales. "You have to think how little money there is in speedway." Steve's invested in revolutionary equipment to help riders recover more quickly and is the second person to have such capability in this country. He stresses it's an investment rather than any desire to compete with his long-time physiotherapy friend and mentor, "I'm not meaning to do Brian out of business but he lives in Ipswich and I live in the Midlands, which is probably better for many riders. I'm treating Nick Morris, Josh Grajczonek and Mark Lemon – so I'm learning Australian!" Steve clearly has no worries about future work recommendations. Catching sight of Craig Cook, Steve greets him warmly, "What brings you here tonight?" Beneath his baseball cap Craig is affable and slightly diffident. " I'm staying in Guildford tonight and riding at Somerset tomorrow. The next day

[3] Along with the chequered flag, Keith subsequently hands these red flags over to incident recorder Derek Turner to take – possibly in Olympic flame fashion – to Cardiff. It's amazing to think how much profit BSI pocket from Cardiff – the jewel of their SGB Series crown – yet they can't even be bothered to invest in a few flags! This isn't a total surprise given their previous lack of investment – for example, in shale covers prior to the aborted meeting at Gelsenkirchen. Judged by the programme, BSI's use of Birmingham speedway's handheld signalling technology at the Cardiff SGP is a source of significant pride at the club. "Birmingham speedway clearly has the best race flags in the sport, as straight after tonight's meeting they'll be taken away for usage at the British Grand Prix. So whenever the chequered flag is shown on Saturday, it's the same one brandished by Keith Kershaw on a weekly basis!"

I'm going to Cardiff. You've got to enjoy yourself, otherwise you don't ride well! I'm going to have the odd drink at Cardiff but I'm not going to go silly. I'm only making up the numbers at Newport the next day. I ride it well. I always get double figures there. It's Leigh Lanham to beat. I'm busy the next week. It's Glasgow Monday, Poole Wednesday, Birmingham Thursday and Edinburgh Friday. It's all gravy – I need the money. I'm skint!" Last season at Workington, I pitched my display table next to Craig Cook's van with a Jack Russell inside acting as guard dog. "She's called Maddie. She's my sister's dog. She was born deaf – the dog, not my sister! My sister's at university in London." Craig is quick to praise the Coventry signing of his Danish ex-Comets teammate. "Peter Kildemand is going to be a star. He's a year or two younger than me. I'm 24 he's 21 or 22. Last year his riding was good but his gating was off, this year he's sorted his gating out."

Over in the staff car park, Ben Barker chats boisterously before flamboyantly accelerating away in his sporty black Peugeot 207. Rick Frost and Julie Mahoney pick their way over the dimly lit uneven car park surface. It's a night that's seen the Panthers gain a valuable away match point. For Julie, like for every other speedway fan in the stadium, the manoeuvre of the night was Nicki Pedersen's.
"I said to Nicki after 'I went hurr!' and he said, 'I went hurr!'"
[Jeff] "That's why you signed him."
[Julie] "When you think it's our first time here we did well."
[Rick] "We would have beaten them if we'd been at full strength."
[Julie] "We only got one point from reserve."

**23rd June Birmingham v. Peterborough (Elite League) 49-44**

# CHAPTER 11

# Newport

## "If speedway was raced over one lap, James White-Williams would win all the races!"

24th June

Gardening

Thank goodness speedway is a summer sport! Oh wait, hold on a moment – there's dark cloud, drizzle and rain ahead of the Newport 'Drillerman' Hornets and Dudley 'Building Society' Heathens National League meeting. It's my fourth meeting in eight days in three different countries. So far, two took place in drizzle/rain with the other postponed. At the start of the traffic cone slalom provided by the road works nearby to Hayley Stadium, a substantial number of ducks waddling across the road cause further delay. Consensus inside the stadium says it's definitely weather more suited to ducks than speedway riders.

Newport club photographer Hywel Lloyd succinctly sums up the metereological dilemma faced by the promotional team of Steve Mallett, Nick Mallett and Peter Mole (along with general manager Laurence Rogers). "The problem is if you open it and it's rained off you look stupid and, if you don't, it'll brighten up! Another thing is that the people who've come here tonight come from all over so, if it is off, they'll all want their money back at the end!" Track curator David Dean temporarily halts his tractor on the dark coloured Hayley Stadium shale surface. Like volunteer speedway staff everywhere, David is open and friendly, "Have you got a new book out?"
"Yes."
"What's it about?
"Speedway!"
"Am I in it this year?"
"I think you're mentioned."
"I was apprentice to the big boss and now I am the big boss!"

Though David enjoys responsibility for track curatorial matters, in the drizzle the big, big, big boss Steve Mallett dashes over with an expression that indicates that his own internal mental landscape suffers from much darker clouds than the jet black ones that hover overhead. "If this is fuckin' off, I think I'll chuck it all in!" My notionally helpful comment, ("It's forecast to be sunny on

Sunday") gets short shrift, "If I have to do the track tomorrow, I'll fuckin' kill myself!" Riders and mechanics from both teams shelter in their respective section of the pits. Tonight's SCB referee is Willie Dishington from Motherwell. This evening, he'll need all his listening and negotiation skills in order to fulfil his responsibilities. Steve Mallett wants to speak to him, "Where's the ref gone?" Sensibly, he's gone for a bite, "He was here but he's gone to McDonald's!"

With the trackshop primped and primed for action but the turnstiles resolutely shut, franchisee Mike Golding sits front row in the grandstand to watch the weather as well as passers-by. He chats amiably with almost everyone. Steve Mallett's mum and Nick Mallett's granny, Jill, practically wearing out the ground in front of us as she regularly strides to and from the pits. Jill has the latest news, "The boys say they're happy to run!" With a hint of drizzle in the air, Mike sniffs the sky, "I think it's stopped." Everyone is a metereological naysayer including the man lingering under the lee of the grandstand roof, "I don't think it's a very good forecast." With years of promotional experience behind him, Mike's vigilant but wary. "I hate cancelling on the forecast. Anita phoned and said it's raining heavily in Somerset (Burnham)." Mike knows his speedway onions and has the connections to match since he's completely plugged into the informal networks of communication that characterise the speedway world. Though he knows the people, information and gossip – he wears his knowledge lightly. "We're getting all the news here."
"You're an authority figure."
"Not here, I'm not. Nowhere any longer! I think this meeting has to be treated different to any other. People are in the area. You can keep them outside until seven but, even if you have to give the money back, you can sell them drinks and burgers – so you might as well make a bit of money!"

People complain about the Malletts but never to their faces. Others judge them by their frequent trade press pronouncements. Mike feels many comments are ill-informed. "To be fair to the Malletts, the track is really good. The shale is smooth. Look at the grader going round! Last year, they had the wrong type of shale. It used to turn to dust almost straight away and, then, whenever we had delays – rider injuries mainly and the County ambulance takes an hour to come on a Sunday – you've lost the track, no matter how much water you put on it! Different shale makes all the difference. The Malletts are very good people who really love their speedway! If you chat to any of the riders, they'll tell you that they get paid every week, which isn't something you can say everywhere! And the Malletts have them over and feed them. Leigh Lanham was asked by a rider thinking about coming here what they were like and, he told him, he had all his money but couldn't say the same about his doubling up from last season. Okay, they are passionate, they can have massive big arguments in front of the crowd but that's just the way they are and how much they care!" Jill's back from the pits again, "Terry says 'can you go on the computer and look at the Cardiff airport weather forecast?'"

The sky certainly isn't clear and the drizzle remains so a "wait and see" policy currently still looks sensible. Mike Golding sees a bright side, "At least they're not all standing on the track looking at it! I used to hate it when people did that." "Packing the shale down?" "Making the water move up more like. It's the last thing you need people to do! We had a meeting abandoned here after four heats and an early season meeting at Glasgow was cancelled because the track hadn't settled. And the other week, our meeting at Berwick was off at 11 and then – the next night – it was off at 6 at Newcastle. So, though we've got a new [Premier League] team, no one has seen what they can do yet! It's probably too late to save the season. They're still an unknown quantity. Steve originally built the team assuming all seven riders would improve their averages but that wasn't going to happen. One or two might – that's what you usually hope – but all seven is very unlikely!"

If you read the *Speedway Star,* though attendances still don't reach the levels the promotion desire, apparently the club enjoys a greater degree of financial stability because of sponsorship from a dentist called Mole. Tonight's National League team are named after said sponsor (Newport 'Drillerman' Hornets). Mike tells me, "Everyone thought the Mole was distantly related to Tony – because of the name and where he comes from – but he's not. I think the granddad John and Jill are driving things now along with the money from the dentist. It keeps them going. There was talk that they'd pull the plug on the Hornets but the dentist has guaranteed the riders' pay until this weekend then, after that, there's only three or four meetings left. Attendances haven't been great, probably 400 for the Wasps and 150 for the Hornets. They're the same people every week so you can't sell them any merchandise – you need new people to buy that! The people here are very friendly – I've made some good friends and they always come in for a chat – but don't buy anything! It's a bit like a crèche in the trackshop sometimes, dads bring the kids and let them go. Anita got loads of knick-knacks they can get for 50p. We don't make any money it but it keeps them amused. I can see how things can be a bit disheartening sometimes but, as a promoter, you can't criticise the crowd or complain to them. You have to go round saying everything's okay in public, even if it isn't!"

No trip to Newport is ever really complete without [the late] Neil Street in the pits advising the younger riders. Mike confirms his notable absence, "Neil didn't come back from Australia after the winter. There was talk of a meeting for him. They announced it here and Jason [Crump] was going to organise the riders. But they couldn't make the date, or something, and it all died away. Neil is unique! He always has top-class mechanical knowledge, expert insight into riding technique."

To see if there is any clear indication of what's going to happen next, Mike strolls down to the pits (around 6.20). He returns quickly. "I don't know what's gonna happen but my sense is that the Dudley riders are less keen than they were! At this level, they're going to want to ride. The ref is in an awkward position. He might be able to get them to promise to start and give it a go but, if one or two fall off, they're going to say it's the track – whether it is or not! The track still looks good and they put more shale down, so it should be fine. I don't think Gary Patchett is super keen but I don't know? [nods towards the figures on the track] They'll wait for the track inspection." In the absence of news, we look out at the track in silence until Mike asks, "Have you been to Mildura, Jeff? If you look at the Leigh Adams meeting DVD you can see a thick cloud of mossies! When we went they said you'll get to do a Mildura wave and we said 'What's that?' But we soon found out [mimes vast arm movements attempting to bat away mosquitoes]!"

Mike is back here again on Sunday – weather permitting – but isn't going to the Millennium Stadium for the SGP. "Anita and I went to the GP last year because we won some tickets in the Barry Briggs raffle. He was told by Benfield or, at least, we had the impression that they'd be good ones, but we were up in the gods opposite the big screen that was broken and the sound was awful last year. So we decided that we wouldn't go this year and save our money instead. We've been to Prague and we're going to go to the meeting at Gorzow because there's the last round of the Under-21 World Championship at Gniezno on the Sunday, so that will be a good weekend."

With time to fritter, we chew the speedway fat. Leicester Lions entry into the Premier League is widely seen as positive. However, their initial team line-up raised eyebrows. With the voice of experience on his side, Mike notes, "Signing Russians is alright until the season starts and then it's a long way to go! Though there's only three tracks racing it makes a big difference. Darkin started off okay at Leicester." We range widely from National League economics ("Lots of standalone clubs have to fill their fixture lists with meetings to pad it out") to a man with his fingers in many speedway promotional pies (including Dudley). "Gary Patchett has said publicly that he can't afford to run Swindon on his own. The talk in speedway circles was that the Russells would drain resources from Swindon and Terry has already announced that he's leaving, hasn't he? If it goes ahead, we're lucky to have this meeting. Stewart Dickson said to me, 'How did the BSPA allow you to run on the same night as the Pairs?' I think that there was an administrative error by one of the girls in the BSPA office. When Glasgow tried to run a meeting, they were told that they couldn't – if you can believe that! But, since we ran last year, the precedent has been set so they can't say no this year really." It's common knowledge that some current and ex-promoters have unpaid bills. Worsening cash flow is an issue for all speedway promoters in an economic recession, particularly those organised on a hand-to-mouth basis. Speedway has its rules and regulations, albeit sometimes applied differentially. For example, some applications for promotional licences are blackballed while similar applications are granted. Speedway appears to lack a 'fit and proper persons' test. "At one point there were three promoters who were bankrupt! I was never asked about that when I was a promoter. The BSPA didn't really want [name redacted], he forged planning permissions and that but now they've decided they want him. Look they're coming across the centre green! Is that good or bad, Jeff?" Mike bustles off to the pits to find out, "It's on! The referee got the riders to promise to give it a go, we'll see if they finish!"

After the turnstiles are opened around 6.45pm, John Jarvis is one of the first arrivals. "They've opened the gates but with this [points heavenward as the rain gets heavier] it all depends on the visibility. They saw the coach of Danish fans outside – on their way to Cardiff – and had to open! There's supposed to be three coaches of Dudley fans coming too! Are you coming to the meeting on Sunday? It's going to be really hot so it'll be dusty." Laurence Rogers walks past with his mobile phone clamped to his ear saying, "It's passing over! There's something here but it's okay!" His confidence is admirable as too is that of the track staff who ladle out more shale onto the second bend. John Jarvis gazes at all the activity approvingly, "They're going to make every effort they can!" Incident recorder Gavin Morrison's good tan is in sharp contrast to the rain. Given the conditions and comparative inexperience the riders, he could get writer's cramp tonight. "They could come off – especially the reserves – in perfect conditions, never mind when it's raining!" Gavin's tan reminds me to (fruitlessly) search for the ice cream van. Always affable race-day maintenance man Peter Brooks stops by, "You've a lovely display of books now. Do you spend all your time writing?" Club announcer Jason Harrold – possibly slightly unnecessarily – advises us, "The action will get underway in about 10 minutes' time without a parade, I'm pretty sure it'll be straight on to the action!" Weirdly, the latest *Speedway Star* contains traffic information from Jason (advising of potential delays in a seven-mile single lane section of road works on the M50). Bristol-based wrestling and speedway fan John Short is pleased just to find the meeting going ahead. "It's good to have the speedway on! The forecast is terrible! Coming over the bridge it was awful and there was 15 minutes of queues at the tolls."

Before we get underway, there's a minute's silence. Mistakenly I assume it's for Matej Fergan but, actually, it's for ex-Newport Wasp Jimmy Gooch who passed away last weekend at the age of 82. The late Mr Gooch top-scored for Newport at Somerton Park in both 1967 and 1968. He's described in the programme – I suspect euphemistically – as "also quite a character too." Newport made additional efforts to promote tonight's meeting to potentially interested patrons including a trip by Laurence Rogers (and his daughter) to the "SGP practice session to leaflet about tonight's meeting and Sunday!" Because of the rain, pretty well every fan sits in the home-straight grandstand. You can have your pick of places elsewhere in the stadium. Apart from two people in hi-vis jackets I take to work here, there are only two hardy fans on the first bend and another three stood by the exit of bend two.

In medium light rain, the meeting gets underway. Or, at least, should do except that Hornets number 2 Tom Young fails to appear on time so is disqualified under the two-minute time allowance. Referee Willie Dishington sounds the claxon for the rerun almost immediately after the previous two-minute warning expires. Newport reserve Richard Andrews gets out on the track earlier than imagined to replace Tom Young. However, the referee illuminates the stop lights rather than allow the riders to come under orders. Club announcer Jason Harrold insistently advises over the stadium speaker system, "Richard Andrews still has to go 15 metres back!" Apparently much more *au fait* with the rules and regulations of the sport, start marshal Ken Petherick twice exasperatedly shouts back at the officials' booth, "He's a reserve!" When the meeting finally gets underway, each rider has a different approach to coping with the inclement weather and damp shale. Jay Herne gates with alacrity, while his rival number 1 Jon Armstrong definitely looks tentative. The two Richards – Andrews and Franklin – vie for position until the fourth bend where the Newport reserve brushes Richard Franklin aside. Since they've just conceded a maximum disadvantage, the Dudley fan behind me understandably pronounces heat 1 as "Crap!" Heat 2 features Danny Stoneman battle manfully for third serenaded throughout by teenybopper-esque screams from the Heathens faithful. Eventually David Gough's retirement gifts him a point. With conditions tricky, it's noticeable that some riders err on the side of uprightness in the corners rather than broadsiding.

Before heat 3, the tractor grades the track while Jason Harrold advises on important speedway meetings this weekend in South Wales, "Tomorrow there's a big event in Cardiff but there is here too on Sunday – the Welsh Open!" If tonight were the first time ever someone saw speedway, they'd find it processional and far from dramatic or fast! That said, the arrival of Ashley Morris onto the damp shale would quickly dispel this impression since heat 3 sees him battle his way through from third. Ashley definitely rides bravely in the wet. His all-action style verges between somewhere powerful and out of control. Every time he exits bend two, he continues to broadside out onto the straight. Though he accounts for Matt Bates on lap two, his last lap blasts round bends three and four look speculative and doomed to failure up right until the moment he shimmies past James White-Williams to win by the proverbial tyre knobble. Ecstatic cheers from a substantial number of Dudley Heathens fans greet his triumph! After a first lap where Kyle Newman pressures Todd Kurtz, normal service resumes with another processional race. Prior to heat 5, the rain gets heavier. Despite the conditions, from gate 4 Jon Armstrong powers round the outside of the field to lead. By the last lap, James White-Williams's lack of power dictates he has to nurse his bike homewards. Unfortunately it's a journey he fails to complete after a fall on the third bend – to loud cheers from the Dudley fans – fractionally after Richard Franklin powers past him. Comfortable in rain or sunshine, Jay Herne wins his second race (heat 6) while his race partner Tom Young fails to finish after a slip and fall prompts his retirement.

An unexplained break before heat 7 sees some riders wander round to bend four while, behind me, Dudley fans chatter. "They said it was only light drizzle." "This is light for Wales." "The riders are walking on the track, this could be bad news!"

Abandonment would be perfectly understandable. Some riders have slithered off their bikes and the rain is heavier. John Short forecasts possible problems ahead, "The readmission should be interesting! I noticed that they've got the old readmission policy in the programme." [1]

Mike Golding's partner Anita arrives during heat 6 to find the racing suspended. Mike ventures out of the trackshop to see what, if anything, is going on. "Did you hear there's a different readmission policy this season? We didn't know because we don't get a programme. If it's rained-off after four heats, they have to pay £4 to get back in at the next meeting. Not so good when you think that those four races could have been in atrocious conditions! It's eight for no refunds now and twelve for the result!" Apparently things aren't exactly rosy at the Super7 PL Pairs event at the Oak Tree Arena tonight according to one fan, "Just got a text from Somerset. It's raining like this but the meeting

[1] In fact, though there is a new readmission policy in 2011, the exact terms remain at the discretion of each club. Page 2 of the Hornets programme is actually compliant with the new regulations. It states: "Rain-off – Refund Policy

In the event of a rain-off your Readmission Ticket will be valid for reduced admission at the ticket price minus £1 per heat for each heat completed up until heat 12 when a result will be called. Concession and child refund reductions are as stated below. These reduced prices are available on any one of the next 3 meetings not including the BSPA Shared Events upon receipt of the re-admission ticket. Those unable to attend within the timescale can apply by post to the Speedway Office.

is delayed and they're watering the track! Mind you at this level, they're not good enough to be in control." The bloke next to him disagrees, "If they don't ride in this, they won't learn!" Another fan arrives late into the stadium, only to find he's missed six heats and the racing is suspended, "I'm pissed off – I left at 1pm and just got here [20.10] and now it's rained-off, probably!" No sooner do these words leave his mouth than racing resumes as the rain "eases". Todd Kurtz wins, David Gough retires and Tom Perry takes it easy at the rear of the race apparently having run out of tear-offs.

Ex-Weymouth official photographer and press officer Julie Martin ventures out from the pits where she chats to Wessex Bus owners Nigel and Marion Charlton. "They were saying it's a bit of a joke. I said as a business person I'd get it on because everyone's here for a good time."
"There's a lot of Dudley fans here."
"I've been reading interesting things about them on the internet. I think that everyone likes to think that their fans are great but there's arseholes at every club."

Throughout this damp meeting, Dudley keep the scores close helped by a torrid heat 8 for Tom Young. First he's disqualified under the time allowance, starts from 15 metres and then retires. The Heathens take full advantage through Richard Franklin and Danny Stoneman with a maximum heat advantage for the overall lead. With her hair newly dyed and cut in an attractive bob, Julie Martin watches Richard Andrews also completely run out of tear-offs, "I'm going to go on Dragon's Den and invent goggles with wipers and then there'll never be another rain-off again!" With James White-Williams and Kyle Newman in rival teams for heat 9, Jules faces a dilemma, "Oh, I dunno who to cheer for here! I'm gonna cheer for whoever's closest to the front!" Though James White-Williams leads, Jules shouts "Go on, Kyle!" about 74 times up until he takes the lead on the last lap. Jules shrugs sympathetically, "If speedway was raced over one lap, James White-Williams would win all the races!"

With the rain back to its default setting of depressingly heavy, the Heathens riders make hard work of track conditions. On the third bend of the second lap of heat 10, Tom Perry falls and retires. On the last lap, in almost exactly the same spot, Ashley Morris falls but gets up to push home slathered in damp shale. He's greeted by ecstatic cheers from the grandstand Heathens faithful. Heat 11 starts eventfully when experienced Jon Armstrong falls more or less straight away from the gate and proceeds to slither to a halt about ten yards from the start line. Since Jon clears the track quickly, the race runs to completion with a Hornets heat advantage to take the score to 35-31. This weather isn't exactly ideal for those camping in the Cardiff area to see the SGP. Jules is one of these happy campers, "I haven't had anything to eat yet. Buzz rang and said he hoped I wasn't drinking otherwise I'd get back to the tent and talk for half an hour and fall straight asleep. It's probably the only time I talk sense. Buzz is coming down with Dougie Wyer's cousin, if his name means anything to you?"

Heat 12 sees second placed James White-Williams fail to complete a lap since he falls alone under no pressure on the third bend to effectively gift Dudley a 5-1. That said, it's impossible to tell which rider is which in their shale-covered helmet colours. Jules shrugs, "It's 'races' in inverted commas! They're doing their best but what can you do?" Understandably, there is another hiatus. Any casual observer can see the track surface provides no traction for experienced or inexperienced riders alike. Dudley Heathens captain Jon Armstrong wanders along from the pits to ostentatiously check the track surface by the start line before he continues to survey the entire length of the home straight. Behind me a fan asks, "What did Armstrong come along for?" Jason Harrold takes the opportunity to again promote Sunday's meeting here, "Always a quality meeting for a quality field! David Howe, Ryan Fisher and all the Newport top men!" Predictably enough, moments later Jason tells us, "The riders said they'd get to heat 12 and the ref has been to the pits and the meeting will draw to a close. We must congratulate the riders!"

Seventy-five minutes of damper than usual National League speedway action ends with honours even. Jason Harrold reminds us, "The bar is still open and Jeff Scott is still there selling his *Bouquet of Shale*." After their obligatory "champions" chant, pretty well all the Dudley Heathens fans – and, therefore, most of the crowd – file out of the stadium en masse. All the riders now need to give their shale-splattered bikes and kevlars a hard scrub.

**24th June Newport v. Dudley (National League) 36-36**

CHAPTER 12

# Cardiff

## "We tried to get away from the Wulfsport look, so we designed our own!"

25th June

Fan zone SGP stall under construction

For the second year running, St Mary's – the Cardiff city centre pedestrian street that runs parallel to the Millennium Stadium – is designated as a "Fan Zone" where the SGP organisers stage an afternoon of pre-meeting 'entertainment'. At 8am, many stall displays are close to completion as is – at the far end of the street – the sound stage for a concert headlined by the Magoos. I've never heard of them but an article in the *Speedway Star* emphasises lead singer Richard ('Webbie') Webb's speedway (and grasstrack) credentials. Webbie rode for Junior Long Eaton for a year and, when the Conference League started, captained Buxton. His career ended through injury in April 1996 after a confrontation with Exeter's metal safety fence (after Plymouth promoter Mike Bowden fell in front of him). Music has been his focus for the last 11 years. [1] [Footnote 1 Richard Webb tells the *Star*, "We played in front of speedway icon/riders such as Bruce Penhall, Sam Ermolenko, Simon Cross, Chris Morton, Peter Collins, Shane Parker, Colin Pratt, Brent Werner and Chris Neath and we've supported acts such as Whitesnake, Stacey Solomon, Human League, Bananarama, Heaven 17, Belinda Carlisle and DJs such as Pat Sharpe, Gareth Wyn and even Danny Dyer. This year, we're playing festivals with the likes of The Beat, The Selector, Wishbone Ash, The Rubettes, Meridian, The Equals, The Wurzels, The Christians, and Geno Washington."] Though BSI relentlessly promote ticket sales for the Cardiff Grand Prix for practically 51 weeks a year: until a few days before the actual event, details about the Fan Zone were notable by their absence. News did leak of a robot appearing at the Fan Zone. Surprisingly, this isn't Kelvin Tatum but a singing one called Titan.

By his stall, Belle Vue co-promoter David Gordon clutches a credit card terminal in the manner of a traffic warden about to issue a ticket. From the giant notice in his booth as well as the impressive brick wall effect stand-up installation, it's obvious David is here to promote the National Speedway Stadium due to be built in Manchester. From the artist's impression drawings, it looks a lavish, incredibly impressive facility. At this stadium, they'll stage Belle Vue meetings and also provide the facilities to aid young rider development in Britain. To mark the launch of the National Speedway

Stadium, fans can take advantage of an introductory offer to "make history" and "be part of it" by buying a limited edition personalised brick or paver. The promotional literature advises said brickwork purchase provides fans with a "once in a lifetime opportunity to be part of the future of speedway!" For just £49 you can add your name and message to a brick displayed on an internal "Wall of Honour" at the stadium or, similarly, you can put your name and message on a paver located outside the stadium on their "Memory Tracks" pavement. According to the promotional blurb, the name and message options are endless (albeit limited by the size of the brick or paver). "Get your family name, company title or even a special memorial commemorative version set in the annals of time by securing your very own piece of speedway history".

On the cusp of a day ahead helping people make history, David Gordon's concerns are much more prosaic – namely, Cardiff City Council's trading regulations. "The silly thing is that we can't sell anything on St Mary's. It's a local bylaw – you can sell on the streets either side but we can't at all!" To avoid possible intervention from council officials upholding this bylaw, David (and colleagues) intend to duck inside the nearby Arcade to furtively conduct credit card transactions in the manner of a well-organised drug dealer. "You can bet some jobsworth will be along to enforce the bylaws. We'll just step away and do the deals over there!" Since it's my first visit to the 'Fan Zone', I ask, "Where's the Fan Zone?" "It's here – they're going to have a giant inflatable arch over there and, down there, is the stage." "It's the Magoo tribute band, isn't it?" David already knows them well, "Funny enough they played at Chris's fiftieth. After they played their set, we said 'Play some other things' but they were honest and said they couldn't! So we said 'Play your set again!' It was good but all knockoff stuff."

Speedway fans wouldn't be speedway fans if they didn't stare a gift horse (and the once-in-a-lifetime chance to be part of history) in the mouth. Sadly, this opportunity to purchase a commemorative brick or paver is no exception. "On the websites – I don't read them but Chris's wife does – all these idiots, using silly names like Mickey Mouse – immediately complained about how expensive the bricks were! They're only £49 for a brick or a paver. We're going to have them on a special wall inside the stadium. Not a real wall but they'll be on a black background that'll make it look like grouting. Anyway, they say the bricks are too expensive. What the goodness do they know? I'd say most of these don't ever go to Belle Vue – never pay to come in – yet they slag off what we're doing on the Belle Vue forum! We get 4,000 visitors a day and about 10 to 12 idiots post rubbish all the time and that's what people read! I say to the webmaster 'Can't we delete the lies' but it's freedom of speech, apparently." Along with haters on the forums, the last winter of Elite League turmoil also leaves scars but – with the National Stadium to be researched, scoped, funded and built – David looks to the future. "There's a bit of a divide between the Elite League and Premier League promoters. We get on with and see the Sheffield lads – Neil and Dave – but we do our thing and they do theirs!"

If the Fan Zone was a well-kept secret, then Nick Barber's Speedway Fayre isn't. Steve Girdwood, the friendly supremo of ReRun Productions, says, "You are a celebrity in the speedway world now your name has appeared in Nick's advert in the *Star*! I said to him you want to attract people to the Fayre not drive them away!" Despite the rain, last night's Premier League Fours meeting at the Oak Tree Arena got Steve's approval. "The racing was excellent last night. I didn't see it because I was in the pits with Kevin Long doing the interviews because we had a cameraman getting all the action!" GRT Media's Bob Tasker also went. "In 35 years of going to speedway, it's the first time I've said 'Would you like to come under my umbrella and watch them water the track?' The track was arid! I don't know why they didn't water and prepare it so we could have good racing and a possible rain-off rather than awful racing and no rain-off!" My stall is sandwiched between ReRun and the Cradley Supporters' Club. Their stall is initially set up and run by Paul and Anita Partridge. Cradley fans since 1970, they're admirers of Bernie Persson. "There's things going on behind the scenes we can't talk about. We were 16 years without speedway, well, without our club! Everyone is so pleased to be back. We're all the old-fashioned fans. We went to Halesowen Festival and, most people who came to our stand, used to go to Cradley but don't anymore. Some didn't know we were at Wolverhampton and some don't want to go there. If we had our own track, we could double the crowd!"

Fully functioning speaker system permitting, Kevin Coombes handles the Master of Ceremonies duties at the Cardiff Speedway Grand Prix. Speedway's presentational equivalent of a cuckoo, Mike Bennett, covets the role apparently under the impression his radioactively dull ramblings would appeal to speedway's biggest live audience. With his Facebook access and familiarity with the speedway grapevine, Johnny Barber is the man to quiz, "It's been a year or so since Mike Bennett campaigned to be centre-green presenter at Cardiff."
[Johnny] "That's died a death."
[Jeff] "He's helping Leicester nowadays."
[Johnny] "Saving Leicester don't you mean? Kevin Long heard from Lasse Bjerre two weeks running asking 'Why aren't you the presenter at King's Lynn?' Kevin posted that, so the first question when Mike Bennett spoke to Lasse was 'What's this you've been saying about me?' Lasse said, 'I dunno.' It's lucky Lasse didn't swear on the mic."
[Bev] "Johnny's taught Lasse a new word now."
[Johnny] "When he's asked about another rider he says, 'cock' now rather than swear. Tell you what, you should speak to Lasse and do a book about what he thinks of the other riders."
[Jeff] "What does he think of them?"
[Johnny] "Mostly he doesn't like them! The Danes hate each other. Charlie Gjedde and Niels-Kristian Iversen hate each other with a real passion. Hans hates Nicki – actually, they all hate Nicki! Lasse did say he did like Claus Vissing but only because he's no good."

Cardiff on SGP day is a handy place to catch up with people you don't see often or to chat to others you do. The Speedway Fayre is no exception. Since the closure of the County Ground, Exeter's Falcons Speedway are sorely missed by many. With his close connections to the club Tony Lethbridge feels this loss keenly. "I'm thinking of doing a history of Exeter Speedway from 64 to the 80s – the Oakes and Mauger years. I also want to do another book of memories about Colin Hill called *Over the Hill*. I was asked by a website – Exeter Memories – to do a list of all the riders who'd rode for Exeter. We got 309 including the second halfers. Then they said, 'Have you got any photos?' Glyn Shailes worked overtime to get them. We're missing eight, I think. Weirdly a rider from 1948 – Harold Smith who rode one meeting for Exeter and one meeting for Hastings and didn't score for either of them – but it's mostly the 80s we can't get them from!"

Jerran Hart's mechanic reveals some trials and tribulations faced by speedway riders during a wet British summer. "Last weekend we had Scunthorpe, that was off and then there was rain at Berwick and then rained-off at Glasgow. The whole weekend cost them £300 in fuel and that and food for me (I don't get paid) – all for nothing!" Arnie Gibbons was at Somerset last night. "It was an impressive feat of refereeing to get 20 heats in before 9 o'clock. The racing was very good. As you know I'm a bit of a stats anorak so I really had to keep my programme dry. I just had to take their word for it that they were the top two teams. You expect the centre-green presenter to be well briefed but he said to Magnus Zetterstrom, 'Are you watching the GP tomorrow?' and he said, 'I'm riding!' 'Oh! Hooray!' he said." Arnie alertly spots ex-stars of the shale. "I'm doing well on my *I Spy Book of Reading Officials*. I saw Ivan Shears, Chris Shears, Pat Bliss and Malcolm Holloway and Tim Sugar today. But there's still no sign of a new track – he said wistfully!" Arnie recently saw John Cooper Clarke. "He seemed to have a lot of *Daily Mail* attitudes, 'That Nigel Farage talks a lot of sense' and he also likes AV not knowing the contradiction that is with his love of Nigel Farage!" Contrary to its title, British Speedway Yearbooks doesn't always get published annually. Needs must, so Arnie's found his own solution, "Now that there is no speedway annual, I do my own! It's surprising to find how inaccurate they really are! Then I pick up my own highlights and have my own reference book so, when someone posts a question on the British Speedway forum, I don't have to search for six hours. It really is all at my fingertips!"

Arnie ponders how I should improve my books. "Maybe you need a different angle? The strength of the books is the unguarded comments! Glyn Taylor's springs to mind." Obsessive speedway fans inevitably bump into each other. "I see Steve and Sarah Miles everywhere. I assume Steve goes to all the Peterborough meetings too! I exchanged plenty of emails with Steve all last winter. I still think that Rick Frost wants to run in the Premier League." Ex-Reading Racers team manager Tim Sugar passes, "I'm a Speedway Grand Prix groupie nowadays! This is my fifth this season. It seems every other weekend we're at one airport or another." Buxton's Joshua Moss is at the Fayre with his mum, Jayne. Wearing the Union flag, he says, "I want Chris Holder to win!" Jayne agrees, "We all want Chris Holder to win!" Keen to eliminate errors in my future books Joshua tells me, "I don't like Josh, I like Joshua." "He answers to Mossy too!"

On the Dudley stall, headless manikins (upper body only) sport Dudley Supporters' Club T-shirts topped off with photo of Nigel Pearson and Charlie Webster stuck where the head should be ("Nigel and Charlie will be special guests later!") The stall also features an eye-catching original Highway Agency road sign that says "Cradley Heath Stadium 1/2 mile". CHSC Chairman Andy Minshall tells me, "One day we're going to put it back up and point it towards the new one!" The stall's not short of helpers, "There's four of us here. Paul, Simon Massey and Will – who's the team manager." On show, Dudley showcase a full range of memorabilia they've specially commissioned to reflect their own unique style and identity. The Heathens anoraks/jackets are particularly distinctive. "We tried to get away from the Wulfsport look, so we designed our own! If you go to Perry Barr, you can probably see them from outer space like the Great Wall of China!"

Workington team manager Tony Jackson enjoys a brief respite from his Comets duties. They're many people's favourites for the Premier League. The speedway world is a small one, especially so in Workington. I greet him with the news, "Your grandmother is Patricia Hamblin's grandmother's sister." Tony shrugs, "You know more than me! They came from Ireland originally I know that." Love of the Comets runs deep across the generations of the Jackson family, "My granddad Mickey Jackson – he's 84 now – took me for the first time in July 1971 against Canterbury (it was 52-26) and I haven't missed one since! I went on for about a year and a half about going. We used to go as a family, back when people went as a family. He joined the Paras – he saw taller men fail the test – and he was posted to India in 1946/47 during the Partition. He was a bus driver for 40 years in Workington, so everyone knows him. He took them to school and then took them to work later. He used to go to Lonsdale Park, aged 9, to see the speedway. The fence used to be at the edge of river and he could swim over and then sneak in underneath. If the tide was in and the water over the fence, you couldn't get in that way!" It's quite a story, "Shouldn't he have been in the Marines then?"

The fundraising Leigh Adams T-shirts on sale at £10 on the Dudley stand prove popular. Similar questions get asked repeatedly ("How is he?"). Simon Massey replies patiently, "I know as much as there is on the internet. I think that there's another press release coming out shortly. We're doing our bit anyway". Heather McNeilly (ex Glasgow Miss Green) is at the Fayre with her father, "Gollob will probably win but I'd like to see Holder or Lindgren. Freddie looked really fast in practice yesterday!" Cardiff attracts a large number of press attendees – although column inches remain in inverse proportion to the size of the media posse. Press accreditation is organised via BSI's Nicola Sands. In some circles, a SGP lanyard and media badge is both status recognition and a prized possession. Veteran reporter (nowadays without portfolio) John Chaplin arrives proudly sporting badge number 194. He bats off my greeting, ("It's access all areas"), with "actually it's not! You can get into the pits till 6 and then they throw you out! They were really keen to do that last year." Worldspeedway.com website founder, Nick Ward is also a regular in the Cardiff SGP press corps. At least, he provides coverage of the Cardiff SGP. He has a website to run and sends live tweets throughout the meeting, so Nick's very au fait with new technology. "Did you see when Lewis tweeted about how to improve? Middlo told him to keep his foot on his footrest. Chris Holder takes his off occasionally but not every bend!" The speedway news aggregation at Nick's worldspeedway.com website is second to none. There are, however, sensitivities. "We have a rival site – speedway365 – they started linking to us for a week and I got an email from Nigel Pearson saying that we were 'passing on their copyrighted material without permission'. They removed us from lots of clubs mailing lists so, suddenly, we weren't getting any press releases anymore! I got my webmaster to sort it out but we lost about a month through that." Despite the oft-claimed success of the Cardiff Speedway Grand Prix, actual speedway coverage in the national or broadcast media is as common as the unicorns in the Fan Zone. Throughout the speedway season, Vic Rigby of *The Sun* publishes regular snippets of British speedway news (and, skilfully, makes the lacklustre sound exciting). Peter Oakes reports intermittently in the *Daily Star* while Ben Findon of the *Daily Telegraph* annually provides welcome but token Cardiff SGP coverage. Ben's speedway memorabilia predilection is for pre-war programmes but he's also an admirer of the riders of yesteryear. "I've just been chatting to Barry Briggs – he's a really charming man!" Ben's often in the Arlington press box even if reports about the Eagles don't grace the sports pages of the *Daily Telegraph*. "Eastbourne is really on an even keel. The attendance figures are good enough and the commercial side is really going well with Chris. There was a time – last June – when things didn't look so good and there was talk that the club might go under!"

GRT's Bob Tasker no longer tries to solve the puzzle about what drives the sales of his products. "I had an 11-year break and, when I came back in 2005, Berwick were champions and, in 2007, they were awful but DVD sales – via the Shielfield Park trackshop – were excellent in 2007! Since then I've given up trying to figure out or care! Here's a sports trivia question for you, 'What's the link between Sky Sports and GRT Media?' In the Berwick team in 1991 we had Kelvin Tatum, Jimmy Nielson, Richard Knight, Scott Robinson and Kevin Little. Kelvin does occasional co-commentary for Sky and Kevin Little does co-commentary for us. See – your books contain tittle-tattle and controversy but now genuine sports trivia! I was chatting to Barry Briggs on the race suits versus tabards debate."

"Is he a big race suit fan then?"

"Er, no! I'm a big fan of Eric Clapton, I've got all his albums, but would I want a pair of his jeans on my wall? I don't think so. If I got a race jacket, I'd put it on my wall with pride but would I put a race suit on the wall? People come up and ask the funniest questions. I had to rescue my wife earlier, as one of the speedway fraternity was asking her who the rider was on the bottom left of our photographic display?"

On the Backtrack stall, their fancy advert displays require Susie to repeatedly close the fire exit (to make sure the room feels like a sauna). Tony McDonald crouches to take photographs of his star turn Shawn Moran. When I take a photo of Tony taking a photo, from the next door stall Peter Ballinger of Clean Cut Sports warns loudly, "Look out, the parasite is here again!" With work to do and customers to serve, I'm unable ascertain how exactly the status of the Backtrack (selling magazines, books and DVDs), Clean Cut (selling DVDs) and Methanol Press (selling books) stalls differ from each other. Is it our modus operandi? Membership of the speedway inner circle? Prompt payment of invoices? Products sold? It's definitely a mystery. Bob Tasker isn't able to cast any light on what might be meant ("subtle as a flying mallet").

Redcar fans Keith and Mandy Mason liked the Pairs experience last night. "There was a big delay between the heats and the semi final. They did propose to run it as one rider from each team but the riders refused saying it was a Pairs meeting! That would be mad!" Keith must have a kung fu background if the ease with which he rips apart my display table is anything to go by. A Newcastle fan at the Oak Tree Arena also praises the event. "The racing was excellent. There was a big delay between the heats and the semi-final. A woman near us fell down and had to get attention from the paramedics. Earlier they were watering it while it was raining. Strangely they had to because it was so dusty!" Pairs referee, Dave Watters, looks back on his own performance with quiet satisfaction. "After five, they needed to water the track but they found the bowser was empty so I just carried on with the heats – bang, bang, bang! We finished the heats by 20.56 but, then, there was a delay. It's a critical time, people don't mind a few minutes but – waiting around then – gets them upset. Some came up to the box but I told them I don't put on the tyres! We'd still finish before 9.30pm!"

As the Fayre draws to a close, Steve Girdwood surveys the scene, "It's ebbed and flowed – it's been empty then full and now it's full again." Johnny Barber is always well informed, "On Thursday, Scott got booed by the Swindon fans at Steve Johnston's farewell – that's a bit harsh!" But, also, sometimes puzzled. "Why on earth has Mike Bennett started a Twitter campaign for Olly Allen to be in the GB World Cup team? If he gates okay – if not, well! He's not in because he's not good enough!"

**25th June Cardiff Speedway Grand Prix**

# CHAPTER 13

# Wolverhampton

## "They're a nice couple – those Ludvigs!"

27th June

Big ice cream

In sultry late June evening sunshine, Wolverhampton sparkles. At Monmore Green Ladbrokes Stadium Wolves race Coventry for the only time (due to an abridged 'B' fixture list) this season in the 2011 Elite League. During the winter it looked likely promoter Chris Van Straaten faced the loss of all his lucrative derby clashes. Unusually, Wolverhampton didn't stage any friendlies against the Bees or draw them in the Cup. Consequently, there's a good degree of regional fervour ahead of tonight's clash – further boosted by the first ever appearance at Monmore of star Russian rider, Emil Sayfutdinov. Given that last season he had Lord Lucan-like qualities, sight of Przemyslaw Pawlicki could also stir curiosity. In addition, May 2nd (versus Lakeside) was the last time a Wolves Elite League home meeting ran to completion. Postponed and abandoned meetings could also mean some pent up demand for Elite League speedway action.

Backtrack supremo and publisher Tony McDonald is in the West Midlands. "I'm on the Shawn Moran tour – Wolves and then Sheffield, Wednesday and Thursday. The Sheffield promotion organised the speaking events there. They suggested it – Neil Machin and, oh, Dave Hoggart. They're putting him up at their cost in a hotel for five nights in George Street when they didn't have to and, when you consider how it ended, that's very kind of them! They're giving half of the £5 admission charge for the evening event back to Shawn, so they couldn't have been more helpful!" In addition to the Shawn Moran DVD, there's also the Alan Carter book. "Susie had to stay behind to do all the publicity this week. It's running three days in the Halifax Courier, there's coverage in the Yorkshire Post and there's some TV. I've never been to a road-racing event. We're doing Mallory Park and Snetterton on two different weekends with crowds of 30,000 and 40,000 expected. One of those is a Legends Day with Kenny Roberts. It's uncharted territory for a speedway book. We missed the British Grand Prix – you can't have everything – but two film companies are interested! One has already been to Alan's house to do some interviews and filming to make a 30-minute trailer. We don't get any money for that but the publicity would be incredible for the book! It might all come off in Hollywood thanks to Bruce Penhall, which would be an irony. I asked Bruce if it would matter that he's not known? And he said 'it's the story, the tragedy that excites them', and that's what matters not who he is!" The excellent Speedway Plus website quotes Tony on the Carter book,

"It could be the biggest selling speedway book of all time – of the twenty-first century for definite!"

The turnstiles have yet to open. Nonetheless there's quite a crowd at gossip central – aka Dave Rattenberry's trackshop. Dave 'The Hat' Hadden is already there, "I'm 14 short of 10,000 caps!" Interestingly, Dave doesn't have the Danny Bird Racing cap I'm wearing today. Dave Rat snatches a copy of *Bouquet of Shale* to look inside, "I want to read what you've written about me!" Finding a section he stars in – until Ludvig Lindgren interrupts – he's briefly the speedway equivalent of a talking book. Ludvig's got an impressive tattoo on his inside right forearm. To the casual observer, Ludvig might have severely limited his future British Speedway career options with a tattoo of what looks to be the Belle Vue Aces club logo. Rat soon has left-handed Ludvig signing a substantial pile of photographs, ("Sign some of those with the Swedish flag too"). Like many of the younger speedway riders with international call plans on their cell phone – plus time to kill at airports and trackside – Ludvig tweets on a regular basis. Many Scandinavian and Nordic riders show better command of English than local riders. Ludvig quickly bats off my praise, ("I've been enjoying your twitters") with a shrug, "Oh yah?" Soon Ludvig retreats back to the pits with borderline writer's cramp. Dave Rattenberry immediately displays the signed photographs and says wistfully, "They're a nice couple – those Ludvigs!"

Tanned from his trip to North America, Bill Gimbeth stands coiled and ready to serve behind the trackshop display counter. "I've been to California on a Travel Plus tour. We saw four meetings and went to Death Valley – twinned with the BSPA Headquarters in Rugby – Alcatraz and the Grand Canyon and we saw the Queen Mary." *Express & Star* newspaper vendor Pete Blackwell parks up to prepare to display tonight's edition and hopes for better weather. "Last Monday it was fine then – at 7.10 – the heavens opened and didn't stop all night. They waited till about quarter to eight but then gave up. I've been coming on a Tuesday now to see the Heathens. I've been impressed! There's some good riders. Sometimes they ride on the edge of their ability. Rome isn't built in a day but you can't tell them!"

Coventry's Colin Pratt wanders past so I embrace my inner Michael Fish, "You've brought the nice weather with you?" Colin isn't up for meteorological discussions, "Don't talk to me about the weather!" With ongoing hunger pangs to satisfy, Dave Rat eats an ice cream boat with gusto. In addition to a substantial quantity of ice cream, it has all the trimmings – flake, nuts and sauce. It's Dave's regular appetiser before the turnstiles open. My query ("How much are they?") doesn't clarify things, "They're free for me."

Dave stops mid-mouthful to ask, "Did you know John Woolridge? You would if you saw him. He died today! I honestly don't know how old he was. Lovely man! He gave up the Stoke team manager's job when it got too much for him. I saw him last week and asked how he was and he said, 'I haven't been so well.' He had a heart attack today." Justifiably regarded as one of speedway's charming men, 72-year-old John enjoyed longstanding links with Wolverhampton Speedway – attending every Monday night – and was a close companion and friend to taciturn Wolverhampton team manager Peter Adams. Never at his most conversational prior to a meeting, tonight Peter looks sad, distant and downcast as he dashes across the stadium car park. [1]

[1] This sad news headlines the meeting report in the next night's edition of the *Express & Star* newspaper: "A tremendous derby match with some wonderful racing was overshadowed by news of the death of John Woolridge. The former Wolves co-promoter and Stoke team manager, a charming, knowledgeable and immensely popular figure in the sport, died at lunchtime yesterday. Monmore Green flags flew at half-mast and a minute's silence was observed before last night's match. And what a clash it was – one which Woolridge would have enjoyed to the utmost. ... Wolves team boss Peter Adams paid an emotional tribute to his great friend John Woolridge after last night's win over Coventry. "He was unquestionably my best mate," said Adams simply. "I've had some dark moments in my time, but he was always there for me – often at times when he could ill afford to be supportive himself. But he always was. I spoke to him today at 11.30, he told me he wasn't feeling very well and he would give it a miss tonight but I would see him next Monday. But obviously I won't. We'd done a million miles travelling together over the last 30 years. I think he was on his hundredth car and put £100,000 into the pockets of Indian restaurant owners as well! He will be sorely missed."

Parked up inside the stadium is a bright yellow coloured three-door hatchback, its bodywork festooned with eye-catching information about Wolverhampton Speedway ("Britain's favourite family sport", action photographs, logos, phone number and website URL). Wolves presenter Shaun Leigh stops when I exclaim, "Oh look at that, that's great!"
[Shaun] "What's that?"
[Jeff] "It looks nice."
[Shaun] "It's a Fiat or something."
[Jeff] "It's an Ibiza. Maybe people think "no brakes, no fear" applies to the car?"
[Shaun] "Probably – look at the number plate [A9 SSL] – simply sliding left!"

Though we're in the football close season, the afterglow of Sheffield United's relegation burns brightly. "It's always good init? It's always fun when your rivals have a dip in form. What a difference a year makes – us now being taken over by Milan Mandaric, being debt free and looking up while they're struggling! Their chairmen were in property – when things were booming, they were laughing and, now it's burst, they're not laughing so much. What's even funnier is that their manager is an ex-Sheffield Wednesday player, Danny Wilson. Even before they unveiled him, there were 500 fans demonstrating outside against it!" Shaun missed the Cardiff SGP, "I've got so many jobs in Sheffield, I couldn't go. I've got half a dozen DJs." Shaun enjoys his work at Scunthorpe and affectionately recalls visits by the Plymouth Devils and their promoter Mike Bowden. "When the Conference League first started it was intense between Scunthorpe and Plymouth to be the top dogs. The first time Bowden walked out – you know how he dresses – drainpipe trousers, black shoes, white socks, baldhead, and long hair. I just thought Max Wall and said, 'Good Evening!' He didn't see the funny side. I primed the crowd the week before they came for the second time telling them, 'Don't forget everybody to wear your white socks next week!' When he came out on parade everyone put their socks up on the fence! Have you been to Plymouth this year? We were going to send him a voucher for grass cutting! They said it could have done with a bit of a cut. It wouldn't be so bad but it's not so big, is it?" Levity quickly turns to sadness, "It's a great shame about John Woolridge."
[Shaun] "What?"
[Jeff] "He died today."
[Shaun] "Bloody hell! Never! I don't like it when people I know die. It's always the nice ones though isn't it?"

Modest speedway reporter Tim Hamblin arrives. He's instantly not pleased by my greeting, "Did you know John Woolridge?"
"I'm not liking the tense you're using."
"I'm afraid he died today."
"He was a travelling companion for Peter Adams. He was knowledgeable, self-deprecating, old-school charming. They'll feel that at Stoke and here and everywhere and throughout the sport! He was one of those pillars of the sport you feel will never be removed, part of the furniture. He'd have loved tonight: sunny day, local derby, the first appearance for Emil – he'd be champing at the bit!"

Redcar fans Keith and Mandy Mason are on tour, "We haven't been back home yet [from Cardiff]. We're working our way back. We were in Newport yesterday and here today." The substantial queue of fans is released into the stadium and rush past the trackshop. Like a Town Crier, Dave Rattenberry tries to entice, "Look, photos from Cardiff! Herbie Hancock! Quick off the mark we are." Visiting from Germany, ultra-keen speedway fan Jurgen Gilsdorf is on yet another British speedway tour. "Cardiff was good. I think the ref did the wrong decision to exclude Sayfutdinov. It was a good crowd – the best."
"It was 1,500 people less."
"Not much, that is a good crowd. Britain has not top riders. Nicholls and Harris not strong enough.

£29 was a good price. It is around 30 Euros. When you go to Copenhagen it's 50 Euros – that is expensive. At Cardiff, you have a seat under the roof and nice facilities. Okay, you are very far from the track – that is a problem. When Belle Vue is ready in 2013, they can stage another GP in the UK. It is the third largest city in the UK with a good airport and roads."

Wolverhampton club historian Mark Sawbridge is heavy-hearted about comparative reserve strengths, "They've got Lambert and Kildemand at reserve, whereas ours aren't up to it! My Coventry friends tell me Emil will gate like a demon and then find himself close to the fence." Luckily there's not just the racing to look forward to, "There's an old rider, Tommy Sweetman from the 60s. He's over from New Zealand. He must be in his 80s. His son, Chris, used to be the mascot. I've been in touch with Chris via Facebook. They emigrated in the 70s and have only been back a few times, so that's exciting." Like everyone else, Mark's shocked at the sad news about John Woolridge, "He had a big heart attack about 20 years ago and another one later but he looked great on it – a nice man!"

Fifteen minutes before tapes up, the very dark clouds overhead start to spit on us. Some Coventry fans complain about the traffic ("Take That are at Villa Park – Junction 46 coming North was very busy"), while Wolves fans sarcastically savour the Bees line up, "I think it's the first time you've had a full team this season!" Over at the trackshop Graham Williams tells me, "It's been bad for weeks, mate! People aren't buying nothing especially straight after Cardiff."
[Bill] "People are slagging off Freddie Lindgren because he didn't do well at Cardiff."
[Graham] "They don't want to buy his T-shirts."
[Bill] "They ask for Tai Woffinden T-shirts instead."

With the riders out on parade and flags at half-mast, the Wolverhampton presentational team strike a genuinely sombre note. With a voice tinged with barely suppressed grief, Peter Morrish passes on the sad news about John Woolridge. Ian 'Porky' Jones puts aside his usual presentation joie de vivre, "It's taken us all aback at Wolverhampton speedway". After an immaculately observed minute's silence, presentational reality returns with the rider introductions. Though professionalism wins the day, clearly many people's minds remain lost elsewhere.

Speedway referee Chris Durno is now a FIM official. Typical of his professionalism, he's already deep into longtrack research ahead of his continental officiating duties. "Longtrack is all about engines, it's all about power!" Chris knew John Woolridge too, "He passed away twice in the changing room at Telford about five years ago. They resuscitated him. He got the bleep and they brought him back round. You can't think of John without thinking of Stoke." Chris defends the Stoke promotion against their internet critics. "Alright there was a two-hour delay at Stoke because he [Dave Tatum] hadn't a paramedic. But it was a late fixture and, because it was an additional fixture insertion, he failed to doubly check to tick all of the regular boxes and forgot the one about ensuring the paramedic was coming."

Sayfutdinov makes an inauspicious Monmore Green debut since he throttles off into the first bend. It's an easy win for Freddie Lindgren. After all the talk of the so-called weaker Wolverhampton reserve partnership of Tyson Burmeister and Ricky Wells, they're untroubled after bend two. Behind me, there's amazement, "Go on – this could be a first!" When I ask, "Who could have predicted that?" Wolverhampton Wanderers football club press officer and Wolverhampton speedway programme occasional contributor, John Hendley exclaims, "Not me!"

Close by is a meek-looking fan with a giant-sized clipboard who – throughout the meeting - ostentatiously explains the intricacies, rules, regulations and tactics to two bemused ladies. "It's a technical thing: when a rider follows home his teammate he gets a bonus point so that was paid 2..." This well-meaning teacher continues at great and confusing length. These explanations are guaranteed to dampen the ardour of even the most curious keen-to-learn speedway newbie.

When I ask the lady (pupil) next to me, "Can you explain that bonus point thing to me again?" her mind must be temporarily numbed by the torrent of insight, since she looks slightly aghast, "Er, no, but he can!" Out by the start line, it appears the recession impacts start-line girl numbers since we have only the added glamour of a solitary start-line operative. Luckily at Monmore, the entertainment is provided between races by 'Porky'. Heat 3 sees two races in one. In the battle of the number 3s, Ryan Fisher bests Ludvig Lindgren for third while, upfront, Przemyslaw Pawlicki and Woffinden race with verve though little overtaking after the second bend.

All the drama in the initial attempt to run heat 4 takes place on the back straight. PK blasts from fourth to second, possibly discombobulating Edward Kennett since he rears magnificently and flings his bike topsy-turveying away. Edward lies prostrate with Wolves fans' cheers echoing in his ears. The rerun sees Kildemand fall and, to widespread amazement, SCB referee Paul Carrington orders a rerun with all three back. It's a reprieve the Dane fails to capitalise upon when he finishes behind Peter Karlsson and race winner Ricky Wells (6 points from two rides).

Between the races, Porky provides his usual trademark avuncular race-day chitter-chatter. I'm not listening but realise that perhaps I should be when club historian Mark Sawbridge asks, "I hope you're getting these bon mots down!" "What did he say?" "I dunno, I dunno, you have to block him out!" John Hendley advises what to listen for later. "Every week – before heat 15 – they play some music by a German composer, oh er, and he says the same thing every week, 'Four men – four machines – one destination!' It's quite something!" Before heat 5, the solitary start girl takes the trouble to clear a discarded tear-off from the track. While we wait for the two start marshals to position the riders, a reasonably strong cool wind rustles the half-mast flags and a bright spotlight effect beam of sunshine breaks through the dark clouds to light the second bend. Whether it's mechanical difficulty or his admiration of the Monmore Green microclimate, Bees number 2 Kenni Larsen fails to beat the two-minute warning and is disqualified. It's a decision greeted warmly by Porky, "Hee, hee, hee!" While we wait for Peter Kildemand to ready himself, with laptop perched on his knees *Express & Star* speedway reporter Hamblin catches up with some electronic race report paperwork. Tim expertly bats off my suggestion that lady luck loves Wolverhampton ("It's all going your way at the moment!") with concern about the intermittent drizzle becoming heavier rain, "Why not put a fresh two minutes on and get on with it?"

Though heat 6 is about to get underway, Sawbridge still savours heat 2, "I think that's the first 5-1 our reserves have done this season!" The bespectacled uber-fan continues his mission to explain, "We're in the left-hand column of the fixed gate positions!" It's as clear as mud! Something that requires no explanation are the thrills provided by Freddie Lindgren and Edward Kennett as they intelligently vie for supremacy. Kennett just fails in a last-gasp dash for the line. Tim Hamblin's appreciative, "I'd like to have seen Edward Kennett come round for a lap of honour. He deserves it! Freddie Lindgren rode him close but left him a line." Author of numerous cunningly strategic trajectories around Monmore, Peter Karlsson isn't quick out of the starts tonight. Nevertheless, he remains cunning as a fox. One moment he appears underpowered and out of contention, the next a last-gasp burst of speed sees him nearly snatch second from Ryan Fisher. Tim's at a loss for an exact explanation, "Maybe he's carrying an injury or something?" With seven races complete in 46 minutes, as the rain spits on we enjoy a track grade. Understandably, the Monmore faithful greet the heat 7 Bees maximum heat advantage with sustained apathy. Porky wonders plaintively, "Are we going to get a Coventry Bees chant?" Missing any reply, Porky mischievously starts his own, "Give me a B!" Porky quickly completes our four-letter word (BEES) spelling lesson and then comes over all philosophical, "What does it all mean?"

With two perfect races already under his belt, Wells blots his copybook with a massive rear on the second bend. Ty Proctor then gets out of shape to gift an untroubled second successive maximum to the Bees. John Hendley's complaint, "We're making a pig's ear of this" finds further proof before heat 9 since Woffinden and Ludvig Lindgren line up at the wrong start gates. Monmore Green track expert (and default guest of choice) Kennett wins easily. Ludvig and Tai comfortably see off the non-existent challenge of Simon Lambert. Amazingly, the speedway expert with the giant clipboard insists the ladies in his class carefully study this masterful exhibition of team riding. "Come on, Luddy! Tai's gonna put him on the front to stop the pass!" With Lambert over half a lap away, it remains a theoretical threat. So much so that Woffinden insouciantly rides the length of the home straight with his front wheel off the shale (albeit as Kennett accelerates away in the distance). By the end of a rain-flecked heat 9, apart from Lambert, Sayfutdinov is the only Coventry Bee without a win or paid win.

Paul Carrington releases the tapes for heat 10, then immediately stops the race. The expert fan talks nonsense but pretends to know exactly what's going on, "If he did it again – blue – he might get a warning or a fine." The rerun then sees Ryan Fisher channel his inner bowling ball at Fredrik Lindgren. Fisher flies from the tapes to blast hard and direct toward the airfence. Lindgren anticipates the manoeuvre, slows and cuts back into the lead leaving Fisher by the airfence placed fourth. The next lap sees Lindgren slow slightly to play bumper cars with Przemyslaw Pawlicki. On the last lap, Pawlicki rides hard up inside Ricky Wells who goes down by the airfence with some drama to ensure the Wolverhampton faithful greet their pantomime villain with boos. Referee Carrington awards the race (a decision that looks a little harsh) as a maximum heat advantage to make a material difference to the scoreline. Heat 11 features a slow-motion half-hearted fall from Larsen who stays down until cheers from the crowd prompt him to clear the track. Bees number 1 Sayfutdinov suddenly looks more comfortable and nearly catches PK from a wide line with subsequent cutback in their race to the line. It's hard to establish whether this is Emil's trackcraft or PK's lack of speed. News that I came via train prompts Mildenhall's Michael Slattery to ask, "How far is the station from here?" Michael's always full of questions, some of them surreal, "Are they letting the airfence down?" Heat 12 sees Coventry on a 5-1 until Fisher's enthusiasm runs him into last place. With dark clouds overhead, spits of rain and a 6-point lead, the Wolverhampton management opt to take the interval. Slattery lurks by my table (kindly on loan from Dave Rattenberry), "The trackshop are very nice to you." Graham Reeve passes with a wave and cheery greeting, "Hello John, how are you?"

The start of heat 13 sees Emil Sayfutdinov clip Freddie Lindgren's back wheel on bend one and, thereby, impede his own progress. We're then treated to a Monmore trackcraft master class from Fredrik Lindgren and Edward Kennett. Edward powers inside Freddie on bend three of the last lap only to see his rival cut back to snatch the lead. With lightning reflexes but also intimate knowledge of track typology, Edward cuts back inside Freddie's cutback to nearly snatch the win. It's a brilliant race. The crowd loudly appreciate the skill of both riders. Przemyslaw Pawlicki leads heat 14 until he rears the length of the back straight and concedes position to race winner Woffinden. The stadium speaker system serenades the riders out for an extremely wet heat 15 with the theme tune from Gladiator. Whether it's conditions or a lack of gating gloves, the Wolves regular heat 15 partnership of Lindgren and PK find themselves at the rear. Tim Hamblin's perplexed, "What happened there?" Fredrik Lindgren soon splits the Bees pair though their heat advantage rewards them with an away league point. I'm surprised Peter Adams favoured PK over Tai for heat 15. Tim explains, "They've used Tai very sparingly in heat 15 – at King's Lynn and I can't recall where else. They prefer PK."

While the crowd ebb away, Tim dashes to the pits for post-meeting rider and team management interviews. "Rosco was upset and thinks the ref had a shocker over the Wells exclusion and the two minutes where, apparently, the ref said 'I didn't think he'd make it in the 30 seconds left!' which begs the question, why not wait the 30 seconds and see? It was a funny meeting, first Wolves were going to win and, then, it was an away win and, then, it was Wolves again. I thought Fredrik Lindgren rode a heroic team-riding race in heat 10 – virtually coming to a stop on bend three to let Wells past, which was why Pawlicki probably hit him. What would have happened if Pawlicki had come off? Peter Adams would be upset if he'd been excluded especially as Harris did it twice at Belle Vue. Any rider could reasonably expect other riders to be going at a certain pace! I also spoke to Ricky and he said if it had happened the other way round, he'd expect to be excluded!" Even with an interval, tonight's meeting took less than two hours to complete. Wolves have won but the death of John Woolridge overshadows everything. "Peter Adams spoke to him at 11.30 and John called to say, 'Pete I'm not feeling too good tonight so I'll give it a miss and see you next week.' Everyone's upset, particularly Pete!"

**27th June Wolverhampton v. Coventry (Elite League) 48-42**

# CHAPTER 14

# Poole

## "We all know the riders will turn up and ride for the goblin's gold"

24th July

Autographs

The *Speedway Star* advertise the FIM Speedway Under-21 World Championship Final (Round 1) with an action shot of Darcy Ward under banner headline in bold letters that asks, "Can Darcy make it three in a row?" On his home track and in fantastic form, the answer must be YES. Darcy competes against a high quality age cohort of talented riders drawn from Sweden, Poland, Denmark, Croatia, Russia, Slovakia, Australia and Great Britain. Many verge on household name status while others – though high scores at Scrabble – are known only to the speedway cognoscenti. They'll definitely test the pronunciation of casual fans and speedway presenters alike.

Travelling to Poole by train, the beauty of the waterfront, the magnificence of the 60s era bus station and the central proximity of the speedway stadium are impossible to ignore. On a warm afternoon with high dark cloud, two hours before the scheduled 4pm start time, music blares out over the stadium car park. Luckily within earshot, the closest buildings belong to the police, fire service and railway rather than residents. For this prestigious event, admission and programme prices have increased. It's roughly £1 a race and £4 for a glossy commemorative programme. Poole co-promoter Giles Hartwell patrols the entrance vestibule area in a proprietary manner to discourage or see off would-be freeloaders. Though the 2011 FIM Speedway Under-21 World Championship is decided across four events, this initial meeting is of such high calibre that it should attract speedway fans from across Europe to Wimborne Road. Starman (as he styles himself on the British Speedway Forum) chats to passers-by and also tries to sell them the latest edition of the *Speedway Star*. "I dunno how big the crowd will be. I won't get to see it but the racing should be good. I expect you'll get a place in the stand, if you want it."

Resident outside the country, tax exile Gordon Pairman retains his long-time relationship with Glasgow speedway but severed his association with Belle Vue some time back. He's now also part of the Poole speedway triumvirate of licensed promoters: Matt Ford, Giles Hartwell and Gordon.

"Usually when the bride gets asked she hums and ahs but I said 'yes!' straightaway when Matt asked me to join. It's the most professionally run speedway club in Britain and the most successful, at least, over the last ten years. People say this and that but Matt is someone who cares for the sport! People say on the internet – and one or two don't give the club a good name – but you can see from the comments how important Poole are by the feelings people express either positive or negative. Whatever Matt wants is what Matt gets. If the rules were made that way, Matt would be delighted but they're not! To be candid, the main reason I'm delighted to be with Matt is that he is honest and you can't say that of everyone in speedway! Not just the promoters! We all know the riders will turn up and ride for the goblin's gold – which disappears as soon as you get there – do it for a bit then go off again leaving a mess. All it does is push up wages at a time when the sport really can't afford it. It really can't! This should be a good meeting. Sure, Josh Auty is injured again and the Russian Tarasenko hasn't got a visa again. We (Matt and I) sat there and said Vaculik, Dudek, big and small Pawlicki, Jepsen Jensen, should all win it and that's before you think of Ward or Dennis Andersson. In practice earlier, two or three of them fell off because they were so keen to find out the traction points for later! You generally don't see that in practice that often but these really want to succeed this afternoon. I don't think there'll be any problems with the weather. The cloud's very good because it means it won't be too hot. See that girl over there – she's a start girl at Belle Vue – nice but not the brightest. When you were there for the 80th anniversary meeting, the lady Mayor went out in her gold chain and she said, 'Who are you?' 'I'm the Mayor', she said. 'What do you do then?' It's not a question a cleverer person would ask, ignoring they probably don't know the answer either. And the Mayor said, 'I'm the most important person in Manchester except for the Queen and she's not here today!' I'm here on holiday with my family this week. We're staying in Sandbanks. It's very nice and there's lots to do in the local area. Usually I fly into Southampton Airport and back out again!" Gordon has his metaphorical finger on the pulse of British Speedway, especially at the Premier League level. "Whatever happened to Kevin Doolan? He must be about 32 now? He's definitely over 30 if he's not 32. Maybe age is catching up with him? When he left Belle Vue for King's Lynn he was quite honest about it, he said he had to maximise his career earnings and he couldn't do that in the Elite League!" "Maybe his bike lift jinxes him?" "He should get it exorcised then."

Next weekend sees the latest instalment of the (Premier League) derby battles between the Glasgow Tigers and Edinburgh Monarchs. Though the meeting is at Ashfield, Gordon is a reluctant forecaster. "I can't predict the Glasgow/Edinburgh meeting. I reckon they might win. You won't mind?" "Why?" "Because you'll be going with the dark side." How do these ideas take hold? "Actually I'm staying with Ian Maclean. Funny enough the Edinburgh fans think I'm a Glasgow fan and the Glasgow fans think I'm an Edinburgh fan." "Probably best that way."

Inside the glass-sided back-straight grandstand, pints in hand a group of blokes watch last-minute track preparations or, perhaps, read the centre-green sponsorship messages innovatively presented on inflatable hoardings. Conversation soon turns to the current state of British speedway, "Kyle Newman got 15 plus 1 at Belle Vue."
"What, Kyle Newman?"
"Yeh. There was something up with the track."
"What, the usual Belle Vue track?"
"Nah, worse than that. Chris Harris can't ride it."
"He's a good little rider."
"Yeh but with good little riders at some point they just seem to lose it somewhere. Whereas, with the Aussie riders, they don't have a gap like ours do! Even Ward, look at him last year – he had a massive year and yet he still had a 6 [average] and now he's back on it again! We should have training tracks."

"But where are you gonna base it?"
"All I'm saying is that those league tracks that aren't covered by the dog tracks, they could have it there, couldn't they?"

The great and the good head down to the pits including Neil Middleditch, Gordon Day, Peterborough promoter Rick Frost and wife Julie, Craig Cook as well as Shane Parker (in an Australian top) deep in mobile conversation stood in the grandstand area that serves as the post-meeting dance floor. Also drinking in the atmosphere is the extremely well-travelled Nigel 'Noddy' Fordham. He has notched up 5,022 speedway meetings! "Look at the management shirt Zielona G ra management team gave me! It's a management shirt! I'm the mate of the bloke who runs it."
"Was the five thousandth meeting there?"
"I forget! Nah, it was at Ipswich. I shall see three of these [Under-21s] finals. I shall miss the second one. You can't miss world finals can you? I don't reckon there'll be many here."

The riders file back and forth from the pits to the track to inspect it. Noddy's in his element. "Alright, Marek!" "There's Martin – Martin!" It's Martin Vaculik, "He's a good rider – I know him from going to Chechnya three or four times a year. He rides in the big league in Poland for Tarnow. There's Chris [Durno], we've been to so many meetings together." Shane Parker passes in animated conversation with Darcy Ward, while Chris Holder ambles along behind in his trademark surfer dude clothing. It's a style he favours on race night and days off. "Alright Shane!? Awright Darcy!? You did well at Lakeside!" Darcy walks on without answer. "Look, they're blowing more of those adverts up on the centre green. They're a pain. They really get in the fuckin' way! Is Auty riding?"
"I don't think so."
"Or Vadim Tarasenko, he hasn't got a visa."

Though Noddy's disposition is as warm as the sun, he's frustrated that the complexities of speedway remain unappreciated by the general public. "That's the worst part; people think you just put 3-2-1 down. I do it properly afterwards [waves programme]. It's a typical BSPA thing, not leaving any room for anything! I think there'll be a lot of dust coming up here this afternoon." Jerran Hart walks by, "He won't do nothing today – he's going the wrong way!" "Why?" "Too big-headed! Doing a Bridger."

The hallowed area of the pits is only accessible to riders, mechanics, press, staff and, of course, the privileged few. The long-time bloke on the gate takes an understated but no less zealous approach to his work, "Are you guarding it like a tiger?" "You have to!" Gordie Day says to an inquisitive fan, "They won't be here, they've all got a FIM jury meeting!" Though Poole perform well on the track, it's off the track cancer news that excites. "I had good news last Thursday, I'm clean!"
"How fantastic! It's a testament to your positivity."
"They say that. I'm delighted!"
"And back where you should be."
[To pits marshal] "I thought they were in a jury meeting? They've got a rider briefing at 3.30. If they don't have it then, they'll be late!"

One of the Swindon track staff (regularly featured in the Blunsdon Blog) stands in prime position for access to the pits, track or centre green. "Here's something for your book, it's the Swindon airfence here today! Ours is FIM homologated, theirs isn't!"

Until recently made redundant Phil Chard was the long-time star reporter of the *Dorset Echo*. "I've known it was coming for about three and a half years. I used to be the deputy editor and when they got rid of someone else from the sports department, I knew I'd be the next because of my wages. They want more football not speedway in the paper. It's the football reporters who cover it here now. They're good and very professional – don't get me wrong, they do a good job but it takes a year or two to really get it. I know the history and I think that helps what you do but no one is irreplaceable so that's life." While we wait for the action, the Wimborne Road sound system occasionally interrupts its continuous diet of pop music with quick two-second bursts of loud heraldic style trumpet fanfares. Perhaps, pre-meeting, they'll stage a joust or an outdoors version of an obscure Shakespeare history play?

Roving speedway reporter Paul Burbidge visits many of the countries that stage high-level speedway. "We all see Darcy here every week though it will be good if someone wants to push him! Martin Vaculik wants to join here. Maciej Janowski says he likes it and there's Pawlicki though, they say, it's Piotr to look out for!" In addition to his work for the *Speedway Star*, Paul populates the BSI Speedway Grand Prix website with its 'news', blogs and insight. "I go to all the GPs and I went to all the World Team Cup meetings except Vojens. The GPs will be back again soon and I'm flying to Australia for a week to see the Aussie championships in the winter. They don't seem to have Mildura this year! I followed your advice and joined all the frequent flier schemes. I'm racking them up on BA and a bit on Virgin but Ryanair don't seem to have one!"

This afternoon I finally get to visit the world famous Poole Speedway Hall of Fame located in the understand bar of the home-straight grandstand. After so much success at the club, wall space is at a premium. There are photographs of trophies galore plus individual Hall of Fame photos of key Poole riders from yesteryear. It's like a mini Who's Who of Speedway. Above the bar serving area, tragically the bitter aftermath of the close season upheaval has seen person or persons unknown graffiti Alun 'Rosco' Rossiter's photo. Though the addition of spectacles and a beard aren't overly dramatic, I still exclaim aloud, "They've graffitied Rosco's photo!" Pausing mid sip, a fan retorts, "Surely not! I can't believe you even said that!" "Said what?" "We don't mention his name here!"

Poole announcer Nigel Leahy has exciting news, "Just to let you know that air horns are half price in the trackshop at £3." This news thrills Steve Miles, "I'll have two!" Ahead of the rider parade, Nigel Leahy issues advice, "Let me remind you of our house rules: no smoking anywhere in the stadium, except for the designated areas. And the other rule, of course, make lots of noise!" Without explanation, he continues, "Darcy has said how much difference your sport makes in terms of miles per hour!" The riders' parade truck is the middle vehicle in a cortège of lorries – emblazoned with the meeting sponsor's logo (Prestige Landscaping & Spas) – with their dumper truck style hydraulics rampant. If you fail to look in the programme, you could be forgiven for thinking the sponsor specialises in lorry hydraulics or the supply of skip-sized coffins. With the riders in position by their bikes, Nigel passes on the metaphorical speedway flame, "Now we turn the meeting over to the control of FIM referee Frank Ziegler and FIM jury president Armando Castagna."

With so many exotic-sounding riders competing, it's something of a pronunciation free-for-all on the terraces and centre green. That said, Nigel Leahy gives us a virtuoso masterclass all afternoon. Heat 1 features Martin "VAT-ZOO-LEAK" He nearly snatches second on the line in a race notable for an astonishingly hard and competitive dive by Dennis Andersson up inside Patryk Dudek. There might be three rounds after this meeting but these boys really mean business from – to (gulp!) echo Kelvin Tatum – the get go! Heat 2 actually features two British riders – Steve Worrall (riding instead of injured Josh Auty) and Joe Haines (for the visa-less Russian Vadim Tarasenko)! Possibly channelling the spirit of Spike Milligan in a sketch about Chinese takeaway menus, ahead of heat 3 Nigel Leahy appreciatively rolls the riders' names around his mouth. "In red there's Alexander LOCK –TY-I-VE" (Oleksandr Loktaev), Darcy Ward, "BAR-TOCK ARSE-LICK" (Bartosz Zmarzlik) and "MECEY" (Maciej) Janowski". However you say their names, sight of Bartosz Zmarzlik delights Reading club historian Arnie Gibbons, "He's one of the six riders I've not seen before! Three of them are Poles: Piotr Pawlicki, Bartosz Zmarzlik and Oskar Fajfer. There's also Michael Jepsen Jensen, Dino Kovacic and Oleksandr Loktaev! I nearly saw him in Denmark but, I think, he had a qualifying meeting so he didn't ride." The battle for second behind Darcy (who wins by over a quarter of a lap) sees a crush at the finish line, "It's a blanket finish down at the far end".

Heat 4 soon finds Oskar Fajfer beached on the second bend after a theatrical fall the referee studiously ignores. Each race (so far) sees hard-charging riders compete aggressively for position. Continuing this tradition, Michael Jepsen Jensen and Dino Kovacic both ride such high wide lines that they strike the airfence (but remain on their bikes). Arnie Gibbons marks his programme with quick, authoritative strokes. After he surveys the tractors out on a quick track grade. "We've already had two overtakes which is the same number we'd have had at Leicester if I'd gone there!" It's rare that run-of-the-mill Elite League meetings have the energy, verve or intensity of the U21s. However, in Nigel Leahy World, the regular Elite League thrashings administered by the Pirates at Wimborne Road seem somehow equivalent, "You can come back on Wednesday to see more of the same!" Arnie ignores this flim-flam and, instead, metaphorically frowns, "Slovakia and Croatia have more qualifiers than Great Britain – that says something, doesn't it? It's lucky there are wild cards." Yesterday Arnie watched the British Under-19s at Rye House. "I had a long evening last night, 3 hours 20 minutes. It was 22 heats so you had the wretched tyre change business after 20 heats. There were lots of reruns – there were ten reruns and two awarded races." "Who was the ref?" "Margaret Vardy. There were three races where the race leader fell completely unassisted! [Lowers voice] I think it was a bit unfair putting dirt on the Rye House track. It was a good open field and a good meeting. It would have been nice if a bit shorter and, yes, there were a number of decisions that Margaret Vardy made that people mentioned but none of them were totally outrageous. There was one where she excluded a rider who was clearly clearing the track and the whole grandstand turned to query the decision. So much so that she got the announcer to explain her decision. I don't think the ref should be having a dialogue with the crowd. The really ridiculous thing was when that Tom Perry fell off on the last lap, she didn't stop the race – so the result was Richard Franklin nearly ran into him! The usual National League aficionados were there so the usual discussions about the state of British speedway were undertaken!" Out on the track there are neither bikes or tractors, "Jepsen Jensen must be particularly tired after heat 4 judged by the amount of time he's been given between races." Heat 5 sees the media pundits' (possible) 'speedway star of the future' Michael Jepsen Jensen lead up until Piotr Pawlicki cuts inside him on the third bend. Martin Vat-Sue-Leak is the next to pass, though Jepsen Jensen manages to bat off the attentions of Bartosz Zmarzlik. Arnie diligently peruses his programme, "I see they can't spell Harkess. If you see Matt Ford's piece in the front of the programme he says 'Harkness' and, if you look [holds open centrefold race card] they can't spell Dennis Andersson either – who's their own rider! It's so hard to get right, it's only in the programme every week!"

Oleksandr Loktaev enjoys a dramatic start to heat 6. He gets totally out of shape, locks up and then straightens for his bike to hammer into the airfence at warp speed. Unfortunately, his high-speed unguided missile collects Simon Gustafsson in a fashion that immediately sickens and raises questions about damage levels. With our view obscured by the punctured airfence and the gaggle of medical personnel, we wait for news from Nigel Leahy, "Oleksandr Lock–Tie–I've is onto his feet – it was a big off! Simon is still getting attention from our medical team." Arnie's a keen student of young speedway rider flesh, "I think it was Quantum where I identified some riders for the future and Oleksandr Loktaev was one of them." Though not disqualified from the rerun, Simon Gustafsson is unable to resume so meeting reserve, Paul Starke, takes to the track. Arnie wonders, "Come on, who's going to cheer for Paul Starke?" "There's some Plymouth fans over there." "I don't understand how Plymouth desperately needs a heat leader and what they do is announce they've signed Gino Franchetti! Given how small Plymouth is, I'm not quite sure it's a track he'll cope with anyway?"

With Paul Starke some distance behind, lap 2 of the rerun sees a side-by-side shoulder-to-shoulder battle between Przemyslaw Pawlicki and Dennis Andersson until the Poole rider powers ferociously up the inside. Though Pawlicki battles back, he falls on the third bend to throw away two points and, instead, gift them to Paul Starke. This prompts Arnie to ask, "Will Starke get more points than Worrall and Haines? Starke is the best young British rider here, so far, so long as he doesn't join Twitter like Bridger!" "Apparently it's his determination to succeed that leads Lewis to comment like that on Twitter." "Succeed in being a tosser you mean!" Darcy Ward manages to win his second race even more imperiously than his first with his margin of victory again about a quarter of a lap. Second-placed Dino Kovacic meets with Arnie's approval. "Given that Kovacic is only 17 and didn't have the benefit of having a whole stadium built for him, like Pavlic, he could be the Croatian number 1." Studying the racecard, Heat 20 already excites Arnie since it pitches Ward against the Pawlicki brothers (and Patryk Dudek).

Maciej Janowski wins the rerun of heat 8 after being warned for moving at the tapes first time out. Before heat 9, Nigel says, "News from the pits: Simon Gustafsson has withdrawn with a rib injury and Oleksandr Loktaev has withdrawn with a wrist injury." Arnie remains phlegmatic, "Oh well, it's going to give Ashley Morris plenty of practice!" At the Norfolk Arena, Mike Bennett repeatedly pleads for fans to contact him on Twitter. Nigel Leahy's not quite so desperate let alone so *au fait* with modern social networking technology. Instead, he we can email him (talkpirates@yahoo.com) if we'd like shout outs during Pirates meetings. I half suspect Nigel's pronunciations might prompt Arnie to email him later, "It's Oleksandr LOK–TEE–EV." Before he can dash off an email on his BlackBerry, Arnie's distracted by a flag on display. "That flag fascinates me!" It mystifies me (the flag in questions says APATOR and underneath that FC LUTON) but intrigues Arnie, "Who are Fleetwood twinned with in Polish speedway?" Heat 9 sees Darcy Ward rest on his laurels to the extent he only wins by a mere 40 yards or so (ahead of Martin Vaculik). Talk of Nigel 'Noddy' Fordham's 5,000 plus meetings prompts Arnie to confess, "I'm afraid I'm only on about 2,200 and something!" Maciej Janowski wins heat 10 to make it eight points from three rides. Though patriotism dictates I support either Joe Haines or Jerran Hart, in heat 11 I prefer Bartosz Zmarzlik, "He sounds like a German foodstuff in my view. Possibly a late night snack." Arnie's thinking food too. "Some sort of sausage, I reckon. There's a Polish rider called Szymon Kielbasa who, when you Google translate it, comes out as Simon Sausage!" [1]

Unusually for an afternoon meeting, the dust isn't bad. Despite the almost ideal conditions, pre-meeting favourite Michael Jepsen Jensen settles for second behind Patryk Dudek in heat 12. Arnie isn't convinced, "On present form, Jepsen might not win a race and he was supposed to be rider to watch!" The stadium is packed with regular speedway attendees. Some are anoraks, others aficionados. Arnie jokes, "Oooh I've just seen my stalker down there – Steve Miles – we see each other so often we debate who is stalking who!" During the interval, Mark Sawbridge stops to chat with Arnie. These acclaimed club historians proceed to have an animated and passionate discussion about speedway governance in the 1960s. They converse in short staccato fashion – diving deep into the nuance of British speedways past. It's a lava flow of words and insight.[2]

The ping-pong of their conversation fascinates but remains hard to follow. "Lord Hartley Shawcross was the chief prosecutor at Nuremberg."
"Initially a Labour MP but then a cross bencher in later life! I have something that will interest you – I have a copy of the Shawcross report!"

[1] Apparently, there are hours of endless fun to be had with Google Translate. You don't need access to a computer to guess what Parsloes1928nearly might mean. Arnie prefers to get his Google translation kicks from 'real' identities. "Russian ice racer Edward Krysov comes out as Edward Rat, Roman Povashzny becomes: 'More important than romance', Ukrainian Stanislav Ogrodnik turns into Stanislav Gardener. In Polish, Dawid Lampart turns into David Leopard and Seb Ulamek becomes Seb Fraction."

[2] Away from the track Mark Sawbridge continues to work on his magnum opus – a history of British speedway, provisionally entitled Shale and Pace. It's a book I'd pay good money to read right now. Mark gives each individual year a headline comment. "1963 – Craven tragedy overshadows all" and "1964 – Speedway's Civil War"

"That was the report that defined how speedway is organised and how the leagues relate to each other."
"There was an amendment in 1987 which, typically, they can't locate!"
"If there's a chance I can have a copy of the Shawcross report I'd be very grateful!"
"From the reports, you'd expect it to be massive but it's only a 10 page report and it's still pretty well the Parker report from the year before."

When we move away from yesteryear, the conversation still ranges widely. I suggest the controversial 8point rule will be abandoned at the next speedway Promoters' Conference – especially since both Matt Ford and Chris Van Straaten will own teams with more than one rider with a ridden average of 8 points or above. Mark responds with practised diplomacy, "I think the 8-point rule was only ever intended to be for one year or, actually, it was never really defined!" Before you know it, the question of guests raises its ugly head. Mark's not a fan, "I've still to hear a convincing explanation for guests!" Arnie notes, "Look at the much hated joker! It's now used in the Danish, Czech and Russian leagues. It was used in Poland for a bit but isn't now and it's never been used in Sweden. The fact that guests haven't spread shows how logically impossible it is to justify."

Since he lives in the Midlands, I assume Mark must have intimate knowledge of the catering arrangements (outlined in an avalanche of press releases and Nigel Pearson tweets) for Golden Hammer next week at Monmore Green. "Is the pork sandwich at the Golden Hammer worth trying?"

"Speaking as a vegetarian, I dunno! People used to say it was the reason to go to Cradley. I should say I'm Jewish as well." Arnie again briefly bathes in the warm, still fresh memories from last night's British Under-19s meeting at Rye House. "At the start of the meeting, anyone could win. Going into heat 20, there were still seven riders who could win!"

When the Under-21 FIM World Championship resumes, it's Vaculik who impresses, particularly the way he bounces off the back-straight wooden fence but stays on his bike. It's an exceptional piece of motorcycling skill. Where Mark is with partner Sally is irritating, "There are two women by me in the stand who go mad with the Polish flag every time that Pawlicki comes out and they're British!" Arnie remains phlegmatic, "I don't have a problem cheering on foreign riders." It's an outlook that makes a virtue of a necessity since almost every meeting at a world level necessarily doesn't include many British riders. The brute reality is that British riders are bit-part players within the senior British (Elite) League and, increasingly so, in the Premier League too. There's drama aplenty on the first lap of heat 14 after Darcy Ward finds himself third so proceeds to treat us the race of the meeting so far. Throughout, race leader Dennis Andersson refuses to be intimidated by the close attentions of his teammate. Their encounter is a test of skill, speed and position they both pass.

Prior to heat 16, Nigel Leahy advises us Clearwater Spas sponsor the race. They also have a slogan "When you-relax-refresh". Philosophically, I'd expect clear water at a spa. Glancing at the back of his programme to check if Clearwater Spas are a sister company of meeting sponsor Prestige Landscaping and Spas, Arnie wonders, "Do you have to have big breasts and wear a bikini to use it?" I'm not sure that Arnie is the best adjudicator on big breasts. Sadly I suspect Prestige Landscaping and Spas "outdoor living superstore" in Poole will be shut by the time the meeting ends. Still, there's always the Bank Holiday sale. Who wouldn't want to avail themselves of equipment that looks part Meccano and part discarded stage set remnant from the 60s version of *Star Trek*? By local standards, Arnie reckons the Under-21s programme isn't salacious. "It was the Poole programme that used to have an advert for a lap dancing club!" Heat 16 ends in drama after Patryk Dudek falls under pressure from Bartosz Zmarzlik on the final corner of the last lap. Zmarzlik's exclusion looks harsh but, then, I don't have FIM referee Frank Ziegler's qualifications (ostentatiously listed as "FMN:DMSB" in the programme). Though the race is awarded, Nigel

Leahy tells us, "We've been given a time of 61.78!" Arnie's immediately sceptical, "Hold on! What? The time of the rider who technically didn't finish?" Not quite quick enough to stop stadium wide scribbling in our programme race cards, Nigel is soon back to say, "We've just been told that time doesn't count!"

Between the races, Arnie flourishes his BlackBerry to check emails, "Oooh, I've just won a Reading Racers sticker on eBay!" As an author and speedway man of the world, Arnie is ideally placed to answer a difficult publishing question, namely, "Can I review books without reading them?" Instantly attuned to the world of speedway cyber trolls Arnie retorts, "Is that another Derek reference? 'I haven't read it but I know its rubbish!'" Joe Haines wins heat 17 after a combative battle with Martin Vaculik. While he carefully annotates his programme, Arnie notes, "Haines has been torn to shreds on the BSF repeatedly over the last 12 months." My ability to interpret the current state of the scorecard falls far short of Mr Gibbons, "Theoretically there could be a three way run-off for riders on 14." For the cookie to crumble this way, Andersson, Janowski and Ward all need race wins. Despite the exciting dénouement in prospect, Arnie feels anti-climax, "The shame is that there's a sense that this is heat 18 of 80!" Dennis Andersson duly wins his last race to finish with 14 points. Much more impressively, Steve Worrall – riding in place of Josh Auty – finishes second! Arnie's ecstatic, "Worrall coming second was pretty outstanding!"

From the apex of bends one and two, off-duty SCB officials Dave Watters and Chris Durno study the race action. Nowadays Chris Durno is also very proudly an FIM official. His first meeting on the bigger/world stage is a Youth Gold Trophy Longtrack Final at Tayac in Southern France. While we wait for the next race, Lewis Bridger walks by oblivious to the crowd. Lewis aggressively prods at his state-of-the-art mobile phone though it's unclear whether he's texting or tweeting his latest plain language (or Herbal Life) communication to his Twitter followers. Maciej 'Magic' Janowski comprehensively bests Jepsen Jensen and Hart in the penultimate heat to assure his place in the 14-point run-off. To join Andersson and Janowski, Darcy Ward needs to beat the Pawlicki brothers in the last race of the afternoon. He resists two laps of early pressure before he too joins the run-off. On the second attempt to stage the run-off, Dennis Andersson retires to reduce the contest to a match race Darcy Ward wins ahead of 'Magic' Janowski. In old money, this would give Darcy Ward a hat trick of World Under-21 titles! Sadly the modern way is – à la the Speedway Grand Prix – to stagger the thrill of world championship competition over many rounds. Injuries, visas and travel schedules permitting, the riders contest subsequent rounds in Holsted, Denmark (August 28th), Pardubice, Czech Republic (October 1st) and Gniezno, Poland (October 9th). Today's FIM referee Frank Ziegler FMN:DMSB won't officiate at any of these subsequent rounds and, after this season, retires from FIM refereeing. He's moving on to become an FIM Jury President as well as being the overseeing President and driving force behind all the UEM track racing events.

**24th July 2011 FIM Under-21 Championship (Round 1) Winner: Darcy Ward**

## CHAPTER 15

# Dudley

## "Imagine if he was to come back wearing the Hammer and Heathen on his chest?"

26th July

Jan O. mobbed

Karl Van Straaten supervises the Monmore Green Stadium entrance gate ahead of the 2011 F.A.S.T. Golden Hammer – the meeting staged by Dudley Heathens Speedway and Dudley Stroke Association. There's already a patient queue of fans waiting to enter. With the track bathed in warm sunshine and the meeting aggressively promoted, Karl expects a big crowd. "We had it a bit earlier last year – ignoring the rain – it wasn't the school holidays then. This year we've got the pork sandwiches van too and the match race so it should be crowded!" Club Chairman, Nigel Pearson relentlessly exploits every avenue of communication to advise fans locally/nationally that the meeting features the cream of the National League plus the speedway comeback of Cradley legend Jan O. Pedersen (in three match races against Wolverhampton's Sam Ermolenko) as well as the aforementioned pork sandwich van. Though it's not a factor at this level, Karl confesses, "I don't like the new silencers."

Over by the sleek glass-fronted home straight grandstand steps, programme seller Keith Fletcher is ably assisted on his stall by his 13-year-old son Connor and, yet to arrive, regular assistant Sarah. We're fan-free but plagued by swarms of flying ants. Colin Pratt roams the stadium keen to obtain his copy of the commemorative programme. "Are you selling programmes? How much are they?" "They're £3. I thought you'd get one for free." "Nah I'll support them!"

It's likely to be a bumper payday for the management team of Nigel Pearson, Chris Van Straaten and Gary Patchett. It also further increases the Dudley Heathens profile and raises awareness of the Dudley Stroke Association. [1] After Colin leaves, Keith nods proudly towards Connor, "He's going to have a go on a speedway bike at Scunthorpe! He hasn't been on a bike before. Where can you

[1] People underestimate the seriousness of strokes and often don't recognise the tell-tale signs. Dudley Stroke Association promote awareness through their F.A.S.T. acronym. FACE. Has their face fallen on one side? Can they smile? ARMS. Can they raise both arms and keep them there? SPEECH. Is their speech slurred? TIME. Time to call 999 if you see any single one of these signs. In addition to this helpful diagnostic template, their advert in the programme reiterates a key message, "Suspect a stroke? Act F.A.S.T. Call 999."

go nowadays and have a try? It's expensive to get a bike but there you can have a try on the 125s, I think it is."
"Will you be going on the big track?"
"The little one."
"If he's good enough, he'll go on the big one."

Programme sales volumes vary, "We didn't sell so many programmes last night for the Eastbourne meeting because it was an insert in the Belle Vue programme. That meeting shouldn't have gone ahead. It was rained off after three heats. Some people will keep their programme from then until it's rearranged." "Are the flying ants here every week?" "No, just tonight!"

Keith regularly enjoys seeing both Wolverhampton and Dudley Speedway and also used to travel to other clubs within the area. "The racing is good at Stoke but I don't go so often with the problems they have. You know, with the start and the electrics!" Wolverhampton *Express & Star* speedway reporter Tim Hamblin is an early arrival. Before getting here, he's already comprehensively researched every conceivable angle and pre-interviewed all the main local riders in the Golden Hammer, including tonight's 1991 World Champion and comeback rider Jan O. Pedersen. Tim's keen to watch his trio of match races with 1993 World Champion Sam Ermolenko. "I interviewed Jan O. Pedersen and, I think, he's just going to see how it goes and find out what level he can really compete at and then ride at that level!"
"Is he serious about coming back?"
"Completely! I suppose if you plan to retire, you can mentally prepare but – if you just stop suddenly and unexpectedly – it's different! Jan told me his way of coping was to have nothing to do with speedway at all. He didn't for many years until he rode some meeting at home and won that. He just had nothing to do with it! When I spoke to Sam last night, he said it would be a typical match race. They'd spend the first race seeing how fast they could still go and, by the third race, they'd really be going for it. Jan is going to find out what level he can really compete at."
"Elite League?"
"Let's not run before he can walk! Imagine if he was to come back wearing the Hammer and Heathen on his chest?"
"What about the National League pay rates of £10 a point?"
"It's not finance for him, it's competing! He's happily married and settled and feels this is unfinished business. He's bought a bike off Pedersen – Nicki Pedersen – and got all the gear!" [2]

The rise of social media continues to allow new ways for fans to interact with the riders. Many riders are on Twitter. Tim believes that the Wolves riders already enjoy close links with the local speedway community. "The Wolves riders are really a friendly bunch. I speak with them all very regularly and I only really follow Fredrik Lindgren, Ludvig Lindgren and Nicolai Klindt on Twitter. I don't have an account to see what they're doing. Freddie's posted a photo of his new telly but usually there's things of interest."
"I like Nicolai Klindt's."
"Bolt 18 is Luke Corbett his mechanic. He's only young. I remember when he first started and they had to get permission for him to leave school early. He always wanted to be a mechanic. It's a family thing. In his first season, if things got a little fraught, Graham Jones would pop over and ask if he wanted a hand. He didn't but the offer was there, which helps his confidence. Nicolai Klindt is always the butt of Pete Adams's speeches. Nicolai is pleasant and articulate. He always asks you to explain a word, or its meaning, if he doesn't know. His English is excellent but he's still keen to learn and improve that just as much as his riding! It's because he can be a little gauche – literal – that Pete Adams mentions him."

[2] Impressively, nearly two decades later he can still fit into his own leathers!

On the programme stall, demand is high. Some fans buy two or three. By the entrance gate, club partner Gary Patchett zealously guards against would-be freeloaders. He ostentatiously checks the guest list or occasionally, if in doubt, nips over to double check and ask people to produce an entrance ticket. Keith Fletcher says, "They don't let hardly anyone in for free here." According to the regular *Express & Star* newspaper vendor Pete Blackwell, demand levels aren't so strong on Dudley Heathens race night. "If it's got Cradley on it, they go mad for it but, apart from that, they don't. The sales of the paper are half what they are on a Monday night." Pete's friend interjects, "They hate Wolverhampton so they just put up with it here because being back is the main thing – the only thing!"

Teacher and speedway fan Richard Dalston is down from Cumbria to see the Golden Hammer. He's got local connections and a love of Cradley Heath. Richard admires the thronged turnstiles. "This is something else. My grandfather was a vicar in Brierley Hill. This is really about Cradley. This will work – we just desperately need our own track! It's one of my two or three meetings a year. The price of Virgin trains from Carlisle! But I have the excuse of visiting my parents in Wolverhampton." Richard's Finnish girlfriend gives him a close interest in riders from Finland. "When Joonas Kylmakorpi won the longtrack, there was quite a bit of coverage in the Finnish press. He's the third Finn to be a world champion. There's an ice speedway racer Jarmo Hirvasoja in the 90s and Jarno Saarnen in the 70s in 250cc road racing. He was leading the world 500cc in 1973, crashed at Monza and was hit from behind and killed instantly. As you know, there are two Finns at Eastbourne this season. Timo Lahti looked good in the World Cup at King's Lynn last year – riding in fourth admittedly!" Like many speedway fans, Richard pays close attention to the frequent rule changes. "Let's hope it's a bit more settled this [close] season. I only know Allen Trump from the auctions where he just outbid everyone." He remains sceptical about setting maximum average limits. "You're always going to get average speedway riders getting averages over 8 because there isn't the top end. Take Joonas as an example. Okay winning the World longtrack is one thing but having a 6-point average – everyone is chasing that type of rider on that average – so he can name his price! And when you think it through, you have averages at this [National League] level! For goodness sake, the National League is supposed to be about development!"

By 7pm, they're down to their last box at the programme stall. Apparently, they printed 750 copies for the Golden Hammer. Outside the Speedway Office vital work goes on, "They've put four loads of boiling hot water down on the step outside the Speedway Office. I think that's where the flying ants are all coming from. Because it's a public area, I reckon they have to. The Clerk of the Course [Kevin Davies] has to do it as part of his duties, I reckon."

Well-travelled speedway fanatic Arnie Gibbons is concerned at news he heard on the radio during his journey to the Midlands. "My trip to Newport looks in doubt on Friday. A lorry has crashed and caught fire in the Bryn Pass tunnel so, unless you go via Monmouth, it'll be impossible to get to." Two days ago Arnie attended the World Under-21s in Poole. "In the race-off, Dennis Andersson fell unaided on the first bend and was lucky to get back in. The ref used the old formula that 'I'd already stopped it for an unsatisfactory start' to avoid making a decision. It's interesting, it was one of those meetings that was somewhat in the middle. On the BSF, it was boring, gate and go or brilliant! There was a divergence too on the crowd numbers. I'd say just under 2,000 but, then, I didn't count. It looks like they've got a big crowd here for a meeting that's got all of the greats in it: like Adam Allott. I've got a sneaking feeling that this will be a bigger crowd than many of the Elite League meetings I've seen over the past few years. There's also no semi-final and final tonight!" With programmes at a premium, Wolverhampton club historian (and Dudley Heathens fan) Mark Sawbridge buys seven. Arnie isn't surprised, "At the first meeting at Leicester, I got six programmes for Bob Cole and I gave them to him and he put them in his bag and then he lost it!" Some distance back in the programme queue, Peterborough uber-fan Steve Miles appears anxious about his memento ("We're down to our last 20 programmes"). Steve affects good cheer, "We're here for the biggest club in Wolverhampton!" Bill Gimbeth defends the honour of Midlands speedway clubs, "They don't get queues like this at Peterborough."

When speedway historians Mark Sawbridge and Arnie Gibbons chew the fat, things get seriously detailed, seriously quickly. Again they talk animatedly. I overhear Arnie say, "There were six factual errors in a 2009 *Speedway Star* article by Richard Frost about Mike Parker." According to Arnie, there is an immutable law behind the scenes, "Ultimately, all the promoters hate each other! It's only a matter of time, if they like each other to start with, they do eventually hate each other! It's only a generalisation, of course, I'm sure that you can find an exception. I think I've been reading too much about Mike Parker recently."

Arnie's comment, "There was a big queue when I came down. It's nearly 7.30 and it's calming down" gets a sceptical reaction from Sarah on the programme stall, "Calming down, is it really? Nigel [Pearson] was amazed we were down to the last box." Almost on the dot of the scheduled start time, the programmes sell out. Arnie knows a solution, "The first person that hasn't got one, I'll send them to Mark Sawbridge!"
[Connor] "First sell out for all season."
[Keith] "The first sell out for Dudley this season."
[Sarah] "I can't believe it! No programmes left, sorry."
[Fan] "Fantastic!"

After a frantic search elsewhere, Tracey finds an additional 14 to sell, "It was 15 but I sold one on the way here."

The first match race of the evening between ex-World Champions Jan O. Pedersen and Sam Ermolenko is about to get underway. Ian 'Porky' Jones snatches a few words with the returning hero, "Jan O. practised last night. He's wearing the same leathers he wore 19 years ago. I was chatting to Jan O. and I said '19 years and you can still get in them!'" "Just about!"

Stood inside the Dudley Heathens stall, Supporters' Club chairman Andy Minshall is in expansive mood, "The crowd is still here for it! It might not have been around for 14 years but it's still been in their minds! Look at them all! You can come on a Monday and we're never far behind on any Tuesday." Clean Cut DVD's supremo Pete Ballinger is on a gantry that overlooks the start line readied for action. "Have you had a pork sandwich?" "I've had one already and I'll be having another later." His wife disagrees, "You'll not!"

Sam Ermolenko wins from the gate in a time of 59.21 seconds. Jan O. looks smooth while Sam appears to hold back slightly. Mark Sawbridge drinks in the atmosphere, "They say speedway is dying." Richie Worrall wins the first race of the Golden Hammer though the majority of eyes are on comeback rider Adam Roynon. He flirts with the bend two safety fence to loud anxious oohs from the crowd. The third lap of heat 2 sees Jack Hargreaves unseat Kyle Newman's replacement Jay Herne. It earns him a disqualification from referee Dave Watters. Tim Hamblin takes an understanding attitude to Hargreaves's manoeuvre, "He just went in a bit deep there." I'm surprised it's a rerun as the riders were on their third lap. Tim isn't, "It looked like the riding order could have changed." Out on the centre green, Porky wonders aloud if there are problems with the stadium speaker system, "Can they hear me in the lower bar?" Mark Sawbridge can hear him, "Is Porky looking down at his programme to announce the line-ups and saying 'there is Mr Jay Herne' and 'there is Mr Jack Hargreaves'? It's not like they're women is it?" Porky attempts to engage with the younger crowd members, "When do you break up for the school holidays? Is it two weeks?" [Fans in unison] "Last week!" Porky quickly disguises his holiday *faux pas* with chatter about Heathens club captain Jon Armstrong who's walking on the tarpaulin-covered surface of the dog track. "Hello, Armo is coming round. He's going to lend me his bike at the end of the season.

Tell you what, he doesn't half look like Nigel Pearson! He's a mini-me version. People say I'm Colin Pratt's lovechild!" For some, the idea of Porky on John Armstrong's bike is an incongruous concept, "What about the suspension? Will it have stabilizers?"

The rerun of Heat 2 sees a previously off the pace Kyle Hughes surge to victory. Porky celebrates an immutable law of the shale, "That's speedway – the rider they call Mr Kyle Hughes takes full advantage of the rerun." Mark Sawbridge asks, "Did you get that Jeff? 'The rider they call Mr Kyle Hughes'!" Announcer Stuart Sargeant interjects, "We've sorted out the bike for you but who is going to fit the stabilizers?" Tim Hamblin nods, "Stuart Sargeant is quite sharp. Porky is a motorcyclist. I think he had a go at Stoke once." The rapid-fire heats continue. Heat 3 sees Richard Franklin try every race position except second, "Franklin went from first to third to fourth to third." Before heat 4, Stuart Sargeant has a mea culpa, "I must apologise to the heat sponsors. Heat 1 was sponsored by Arthur Edge who remembers Bernt Persson; heat 2 is sponsored by Attwood Garden Services; heat 3 by Bernard Fogarty who's been a fan since the Summer of '59 and heat 4 will be sponsored by C. Beech & Sons (Netherton) Limited." Such is the commercial attractiveness of the revitalised Dudley Heathens every heat tonight is sponsored. Todd Kurtz wins heat 4 by a quarter of a lap ahead of the diminutive Jon Armstrong. Stuart Sargeant bigs-up my latest book *Bouquet of Shale*. Porky retorts, "That's a damned good read – but you can read it in 30 minutes, you're so clever!" While we wait for heat 5, Colin Pratt manoeuvres through the gap in the fence that separates the dog track from the speedway track. Porky greets him enthusiastically, "Come on Colin, can you get through that gap? Here he comes! It's my pork sandwich provider, Colin Pratt. You're my dad, I've told them." As normal, Colin is a diffident, reluctant interviewee speaking quietly as if testing the reception of a new hearing aid. Colin is impressed with the love Cradley/Dudley (still) inspires, "Fantastic crowd, terrific!"

Driven onward by repeated gasps of encouragement ("Go on Adam!"), Adam Roynon wins heat 5. There's a photo finish for third place at the end of heat 6. When I ask, "Did Brendan Johnson get third on the line?" Tim Hamblin retorts, "Or Herne lose it?" With behind-the-scenes access, Tim is a discrete but shrewd judge of speedway riders' characters. "I was impressed with Brendan's two meetings for Dudley. I was interviewing him in the pits and he was talking about fighting for his place at the time Richard Franklin was doing his post -meeting practice laps. They were always going to bring him back in, so it was nice Newport came in for a team place for Brendan as it worked out well for everyone!"

I'm on the terraces with Mark Sawbridge, Arnie Gibbons and Tim Hamblin, My cup completely overflows with the surprise arrival of Allan Melville, "Speedway's number 1 speedway website told me you'd be here!" Allan might be modest but, since its inception, the Speedway Plus website has become an incredible resource that mixes commentary, history, comment, reviews as well as 360 degree insight and analysis into speedway/speedway tracks throughout the world. This week Allan works in Birmingham. He had trouble getting here because of the crowd, "I had to stop the taxi half a mile from the stadium". Darryl Ritchings impresses Tim with a last-gasp past bend overtake to snatch third on the run-in to the line. "That's a gutsy ride from a 16 year old! Hargreaves did leave him a line but not much of one." Without a programme, Allan Melville struggles to get a handle on the action, "I'm a bit lost, they've sold out of programmes when I got here."
[Jeff] "Mark Sawbridge bought seven of them."
[Allan] "Flippin' Mark!"

Mark Sawbridge's Poole-supporting partner Sally Knight isn't surprised at graffiti damage to Alan 'Rosco' Rossiter's photo in the Wimborne Road Hall of Fame. "Police have narrowed it down to 2,000 suspects prepared to admit they did it!" Ever helpful, Tim explains Sally Knight's status, "She's officially the second item to be rescued in a fire by Mark after the Finn Thomson race jacket." Sam Ermolenko wins the second match race of the night by an even greater distance (quarter of a lap). Arnie Gibbons exclaims, "Oooh, Sam was faster than Tom Perry." Heat 9 sees Jay Herne battle lustily with returnee Adam Roynon. Kyle Hughes wins heat 10 ahead of Jon Armstrong who's serenaded throughout with shouts from Arnie, "Go on Armo!" Arnie wonders if the Cradley fans have put their collective best foot forward with their civic visitors. "It doesn't say much for the brightness of the Dudley speedway fans that they booed the Cradley councillors – who they are trying to persuade with the force of their argument!" Adam Allott wins an exciting heat 11. Arnie says, "Adam 'I finish where I gate' Allott" Tim Hamblin suggests that Adam isn't the only exponent of this dark art, "Like Chris Louis in the Olympique 30303." Allan Melville leaves us, "I'm gonna watch the speedway here from a different angle." Stuart Sargeant reminds the younger fans sent giddy by the school holidays that they too have responsibilities, "Can the children please keep their legs on the grandstand side of the fence – otherwise the meeting can't continue!" Mark Sawbridge tuts, "Is that one of those many namby-pamby health and safety warnings? How many times do the bikes make it to that fence?"

To widespread adulation, Sam Ermolenko and Jan O. Pedersen walk from the pits to the Dudley Heathens Speedway Supporters' stall. Jan O. cuts a dapper figure. He's tanned enough to be mistaken for Jon Cook's brother. Long-time speedway fan, Max Cartwright dashes from his favoured spot on the grandstand to try to meet his idol. "He's off to get his hat signed. He went to the first ever Cradley meeting but didn't have any money to buy a programme. Imagine how much that would be worth now? It's the hat he bought the night Jan O. became World Champion."
"What year would that be?"
"I'm not sure, I think it was 1991. We'll have to have a look at the hat when he comes back!"

Max displays sharp elbows to battle through the substantial throng around the two ex-World Champions. "Max has got his hat signed. He was stuck in a massive queue and Nigel Pearson took the hat off him and got it signed. He's absolutely made up! The hat is actually from the 1987 world final." There's so much demand for autographs that even Nigel Pearson signs a few.

After the interval, Byron Bekker adds initial drama to heat 13 when he flings his bike away. It's the speedway equivalent of a riderless horse in the Grand National since Byron's bike goes much quicker without him. While we wait for the rerun, the crowd start a brief Mexican wave. Arnie wonders, "How can you do a Mexican wave with only one grandstand?" Almost telepathically, Stuart Sargeant responds over the stadium loudspeakers, "That was the Porky Tsunami not the Mexican wave." Apparently not factoring in that his repartee might be a contributory factor, Porky asks the crowd, "Have you noticed the security man at the front never smiles?" Adam Roynon gets disqualified from the rerun so, instead, we see a match race that Brendan Johnson wins ahead of meeting reserve Adam Portwood. Tim Hamblin has curatorial matters on his mind, "I was talking to the man from Breedon Aggregates at Leicester and he said they supply quite a few tracks including Poole but don't go down there much!" Arnie's feeling devilish tonight, "I wonder how this compares with the Isle of Wight versus Stoke crowd tonight?"

On the third lap of heat 14, Tom Perry tries for the proverbial non-existent back-straight gap that – it transpires – is actually non-existent. He backs off and regroups to snatch third on the line from Richard Franklin. In the next race, Ashley Morris bests Kyle Hughes on the line though the optical illusion of our vantage point requires the referee adjudicate ("It's deceptive, though Hughes was going faster at the line!") With time at a premium, referee Dave Watters awards heat 16 after race leader Richie Worrall smashes into the first-bend airfence on the second lap 2. Jan O. Pedersen's back passes its examination so, if he wants, Jan can resume his speedway infatuation. The third ex-World Champion's match race (at 21.40) is, again, won easily by Sam Ermolenko. Porky reminds us (again), "There will be an auction of Sam and Jan O.'s race jackets in the bar afterwards" before he compliments Jan O., "You ride it better than Nicki Pedersen anyway." Apropos of nothing, Porky confesses, "I'm sorry I'm not one of the girls with the boobs off Sky TV!" This befuddles Arnie, "Haven't they all got boobs?"

Jan O.'s clearly a shrewd judge – both on and off the track – since he immediately gives Porky his bike to ride! Porky rides with some grace at quite a lick. "I can do some talking now. I just wanna say what a fantastic night. It's 19 years since I was on the track. It's very emotional! It is absolutely amazingly emotional for me. All I'm doing now is just enjoying life and this is all about enjoying life and this is really enjoying life!" Porky's so keen to snatch the microphone back, he only completes a single lap. Diplomatically Jan O. refuses to reveal any future team or league details, "I'll see how far it goes and see who comes up with the best contract." With all the talk of coming back out of retirement Porky asks Sam, "Will you be coming back?"
[Sam] "Excuse me, have you ever seen my legs? I don't want the left to look as bad as the other one!" Sam soon resumes his standard affable factory setting, "He told me to go full out because he wanted to see."
[Porky] "If you want any tips, I'll teach you."
[Jan O.] "Do I need to gain weight as well? [pause] Have you all had fun out there? All I want to see is Cradley back with their own track!" [3] [Footnote 2 Jan O. Pedersen would make a brief comeback riding two meetings in the Danish League in September 2011.]

Stuart Sargeant interjects, "Great to hear two great professionals on the microphone!" Porky's always reluctant to concede centre stage for too long, "Nicki Pedersen wouldn't let me do that on his bike! I've got nothing left on my shoe now!"

With only four races to go and rumour of a 10 o'clock curfew, we expect and get quick fire heats. Porky reads the line-up for heat 18 in the style of a difficult eye test where each letter recognised is a profound revelation. The race only lasts as far as the first corner of the second lap because Shane Hazelden falls. He then struggles helplessly on the shale like an overturned beetle. Adam Allott wins the rerun for 12 points from 5 rides. Tim Hamblin's anxious to get down to the pits, "I'd like to speak to Adam Roynon but he may have gone before I get down there. Before he was injured – in the space of four days – he'd won in all three leagues the week before!" The last two laps of heat 19 showcase Todd Kurtz and Darryl Ritchings locked in neck-and-neck battle (before Kurtz wins). Prior to heat 20, Arnie Gibbons demonstrates lightning-fast mental calculation. "They could have two run-offs if Worrall wins and Hughes is third – they'll need it for first and third place." It's a doomsday-time-curfew scenario blown off course by a win for Kyle Hughes (ahead of Richie Worrall) to give him 13 points and the Golden Hammer trophy! Though Adam Allott and Todd Kurtz tie, Stuart Sargeant informs us, "There won't be a run-off for second because Adam Allott is second by virtue of race wins."

From the centre green, Nigel Pearson receives the adulation of the Dudley faithful, "I just want to say thanks to all of you who come every week, from the bottom of our hearts!" Steve Miles enjoys the fanaticism of the Dudley fans, "This is the only club that could go straight from the National League to the Elite League!" Dudley Supporters' Club chairman Andy Minshall celebrates and participates in the passion, "We've still got the spirit of the 80s!" Back from the pits, Tim Hamblin reports, "If it had been a week later Adam Roynon might have won and Kyle Newman would have possibly too. Adam is really annoyed with himself. He wanted five races to see if his leg was okay and, instead, he started racing the races rather than racing the track thinking he could win. He has trapped a nerve in his back and shoulder but he's expecting the physio to clear it as he has a busy schedule on Thursday, Friday and Saturday. He's an articulate, engaging and open young man. Obviously in pain, but he still did the piece!"

**26th July Golden Hammer Winner: Kyle Hughes**

# CHAPTER 16

# Redcar

## "He'd fall off so he could be first in the bar"

28th July

Sylvia Morgan & big bear

Half way on the 64 bus from Middlesbrough city centre to South Tees Motor Sports Park, an early-bird Redcar Bears fan flags it down. He's arguably typical of speedway's demographic. Aged late 50s to early 60s in a Redcar Bears anorak festooned with enough badges to gladden the heart of trackshop vendor Dave Rattenberry. The late afternoon sunshine that bathed the Costa del Teesside is replaced by a cool wind. The clipboard lady by the imaginary security gates on the second bend has me on her list. "I knew you'd spell it with a "J" when I wrote down Geoff." Our talk soon turns to the weather, "When I got here at five you could see those hills and, now, they're covered in mist. It's forecast for rain further up the coast but I don't think it's gonna come here. I just got a text from Redcar saying 'what's the weather like?' 'Cloudy,' I said." The stadium is presently fan free while stadium staff diligently prepare their respective areas of responsibility. Since my last visit, an impressive – in terms of size and utility – covered home-straight grandstand now overlooks the finish line. In the programme booth adjacent to the entrance turnstiles kiosk, Janice Jackson industriously unpacks tonight's match programmes from their boxes. Her partner in crime Claire James now works elsewhere, "We've been split up now. I'm still in here and she does the entrance booth." Though many fancy Workington to lift the Premier League title, Janice expects Redcar to see them off at the STMP tonight.

Inside the trackshop, pits gate operative Jim Elsdon breaks off his conversation with Bill Gimbeth. "It's me who should be signing the books not you!" Jim stars in two chapters from *Bouquet of Shale*. Though it will soon be a year ago, totally understandably he hasn't recovered from the death of his sister, Julie Lewis. "You have to get on because life goes on. We've both been involved for 50 years. I'm going to have a memorial meeting next month. It will be 15 heats and 12 riders. There'll be a big trophy that stays here and a little one for the winning rider to take home. I've got the body colours done already. It'll have red and white checks like it used to be on the top and her name in each corner and the memorial meeting at the bottom."

Dave Rattenberry is absent without permission from the trackshop. Actually he's deep in conversation with Claire and Janice by the programme booth. Though some distance away, Rat easily spots Malcolm Wright walking towards us. "It's Malcolm from Sheffield. He used to own Sheffield – he's a millionaire.
[Rat] "Who's at Sheffield tonight?"
[Malcolm] "Rye House. They don't travel well and they certainly don't travel well to Sheffield. I've seen Workington a few times this season and they've been good to watch. Who's riding in place of James Wright?"
[Rat] "Kevin Doolan."
[Malcolm] "It was 17 when I left York. I had the roof down and shorts on and now look at it."

Back inside the stadium clubhouse, Mrs Havelock – Brian's wife and Gary's mum – Marge kindly finds a table for me underneath the photographic display from 1985, when her son was a teenager. "You can sit outside my office."
"That's a great photo of Gary."
"Yes, he was 16 then."
"How old is he now?"
"42."

Promoter Brian Havelock's sits on the other side of the clubhouse eating his pre-meeting pie and peas with gusto. Redcar Bears club director Eric Gladders stop by at my table with a mild complaint, "Ee, you had me down as a Havelock in your last book!"
[Jeff] "Did I? Everyone is a Havelock, aren't they?"
[Eric] "They're not! I'm not!"
[Brian] "Don't have him as a Havelock!"

The clubhouse fills quickly with fans keen for a pint or some food. Brian Havelock sits with some Workington fans and sups a half. Supervised by team manager Tony Jackson, Workington Comets riders industriously kick the shale surface in the vicinity of the start gate. Tony stands out because of his age, his lack of shorts and peaked cap as well as his on-message (Workington Comets) anorak. Climbing the steep stairs up to the referee and announcer's box, club presenter Keith McGhie glances briefly at the pre-prepared sheet I hand him about my book. "I've already made a note that you were coming. Where are you?"
"By the Speedway Office."
[points below him] "This is the Speedway Office."
"In the bar, where Mrs Havelock is."
"Oh, you mean the bar office!"

Amy Winehouse songs blast out in tribute over the stadium speakers. Just as she choruses, "We only said goodbye", I step into the Speedway Office. On race night, you'd usually find Julie Lewis sat beside Sylvia Morgan. Sylvia now works alone with her back to the window that overlooks the start. There's a large sized bear in the chair that Julie used to occupy. "Feel the weight of it."
"That's heavy. What's he called?"
"He hasn't got a name – we should give him one and a body colour! He brings us luck. A little boy comes in and gives him a different name every week. He was Mr Brown Bear and, last week, it was – oh, I forget. It's a nice bear and one of the fans said could we do something with it since his wife died of cancer. He's 86."
"You can see it's an old one from the eyes."

Clerk of the course Peter Morgan (Sylvia's husband) pops his head in the doorway, "Have you seen Brian?"
"He's in the bar."

"That's strange – usually he's by the turnstiles."

Workington programme editor Barry Stephenson warns the Workington fans around him, "Don't say anything or it'll be in the book!" The weather – rather than rider form – is the key topic of conversation. "It's been raining all day in Workington, it was raining when we left and it'll be raining when we get back." Changing topic I say, "Your riders looked confident on the track walk." Barry shrugs, "What all six of them?" Tonight the Comets run rider replacement for injured high-flier Peter Kildemand and have guest Kevin Doolan for their number 1 James Wright.

A red-and-white plastic chain barrier separates the sponsors' hospitality section of the clubhouse from the general public. The hospitality race night host is Steve Harland. Back in 2003, along with Jason Pipe, Steve got together a junior team to race under the Middlesbrough Bears name. Nowadays the club runs again but back then (though the affection was strong in the city and surrounding areas) they were a club in name only. That's the past and Steve prefers to live in the present, "Me wife got your book for me for my 50th."
"You don't look 50! When was your birthday?"
"July 3rd. I'll come and chat later, I'm a bit late and I've got to look after the sponsors."
"The big question – are the lovely pies here?" [1]
"They're not and he's not on the guest list."

In the trackshop Dave Rattenberry, Bill Gimbeth and Malcolm Wright are deep in what must be men's conversation, given how abruptly they break it off. "Can I look in the *Speedway Star* for a minute? I just want to find out what's really happening." With a thin fixed smile, Malcolm exhales, "Um." Redcar presenter Gareth Rogers briefly stops in. He's only fully involved at Redcar nowadays. "I've officially left my duties at Newport to get on with writing my books. It took up two days a week and then there was the travel and phone calls. It was a lot of time that I could use elsewhere."

Sunderland-born Bears fan Brian Swalwell proudly shows me a 7-inch disc from 1974 issued by Decca. It features speedway riders singing 'Rivals'. Among others these include Graham Stapleton, Terry Betts, John Louis, Malcolm Brown, Bert Harkins, Peter Collins, Martin Ashby, Jim McMillan, Dag Lovaas, Scott Autrey, Nigel Boocock and Pete Smith. The all-important question is, "What does it sound like?" Brian undersells the aural delight, "It sounds just like a choir band! It starts off with motorbikes revving up and, then, they sing together 'Speedway! Speedway!' and, then, there's the sound of them riding round the track. I gave it to them in the box – they might play it next week. I can't remember how much it was then but how much do you think it would be worth now?" To my mind, it's priceless. There's also a B-side 'Hoskins still rides' written by someone called Blore (produced by unlikely sounding Luther Birdface for KG Productions), "It's an instrumental – there's no singing!"

Keith and Mandy Mason from Norton-on-Tees nowadays watch close to the clubhouse, "You get too much shale and dust where we used to be." Riders wave from the circling parade truck while 'The boys are back in town' plays loudly over the stadium speakers. After the rider introduction formalities conclude, announcer Keith McGhie says, "Redcar management and fans would like to pass on their condolences to the well-known Workington supporter Mel Pettit who sadly passed away."

Prior to heat 1, 'Let me entertain you' serenades the riders as they dig the shale for real or imaginary advantage. Both Redcar riders – Matej Kus and Gary Havelock – fly away to an easy win with Comets guest Kevin Doolan last. Dave Rattenberry's disappeared from the trackshop, "Where is Rat? Is he getting some food?"

[1] Made by A P Jackson Butchers located at 10 High Street, Ruswarp, Whitby.

[Bill] "Nah, he's in the car park, wheeler-dealering about something."
[Jeff] "He's not paying for sex, is he?"
[Bill] "Ha! Ha! I don't think so."

Keith McGhie highlights the threat of the Comets reserve partnership – particularly Gary Irving, "another ex-Redcar cub!" Keith also welcomes back the rider who's about to finish last, "Making his comeback for the Bears tonight – Adam Roynon".
[Gareth] "He did actually ride at Wolves on Tuesday."
[Keith] "I did say in Redcar colours making his debut."

The impressively fast and stylish Kenny Ingalls romps home with ease. Jason Lyons wins the next ahead of a fast-looking Tomas Topinka – now fully recovered from a bizarre injury sustained last month after a pre-meeting track (!) collision with a mechanic. Though the mechanic escaped comparatively unharmed, Tomas suffered three weeks off with concussion. Dave Rattenberry's keen to establish my sleeping arrangements, "Where are you staying tonight?"
"The Travelodge."
"We stayed there last week. It was too expensive this week – £49! We're staying at the Premier Inn – it's much more upmarket – for £29!" [2]

Though Keith McGhie's dulcet tones sound clearly over the impressive STMP stadium speakers, Gareth Rogers struggles to really be audible. It's a presentational quirk that plagues the whole evening. Keith McGhie metaphorically scratches his head and calls upon his own expertise, "I think you're right behind the starting gate, Gareth. That could explain it? Though, I couldn't guarantee it!" Prior to heat 5, ex-Sheffield Tigers number 1 Sean Wilson dashes into the clubhouse to solicitously apply a damp paper towel (supplied by the catering ladies) to the cut knee of a young girl, "You're as bad as me!" With the meeting well and truly underway, Rat's able to fully indulge his love of speedway catering. He holds his paper plate – brimming with pie, bright green mushy peas and oodles of gravy – with the triumphant pride of an Olympic torchbearer. Despite myself, I have to admit, "It looks nice!" The lady behind the counter quickly disappoints, "They've sold out. Dave got the last one." Mid-mouthful, Rat stops to smile broadly "I did!" Meal soon over, Dave Rattenberry tries to find out the likely shape of the rest of the season (now that team line-ups for the second stage are agreed), "What are the fixtures?" With an exasperated wave of his programme, Brian Havelock exclaims, "I can't tell you! They chop and change. They're out in draft format."

Before heat 6 gets underway, we enjoy a two-tractor track grade. Keith launches a public appeal, "A mobile phone has been found on the ground." Looking set fair for a 5-1 from Havelock and Kus, Rusty Harrison is urged onwards by a passionate Comets fan endlessly shouting, "Go on Rusty, come on lad!" He's apparently resigned to third until a late outside blast snatches second on the line from a somnambulant Matej Kus. Clutching and/or marking his programme like speedway fans everywhere, promoter Brian Havelock extemporises to listeners unknown, "He fell asleep!" Keith McGhie prefers to accentuate the positive, "coming through in extraordinary fashion on the last lap". Immediately Brian determinedly heads off towards the pits, possibly to apply a more hands-on approach denied him within the confines of the Bears clubhouse.

[2] When I return after the meeting, Middlesbrough-born night porter Jamie tells me, "Middlesbrough is a ghost town! Over the last five years, it's got worse. I can say that as I come from here. Everything's shutting down. You go round the town centre and there's shops shut all over. Even the chains are leaving. Its shop – shutters – shop – three shutters. Even when they built the Riverside, they used German steel! It says something about our prices. There's no nightclubs here apart from the Cesspit with the metal detectors we have. The police are always there and it attracts, well, interesting people! When guests come back at two or three in the morning, they get annoyed with me because I come from here but Middlesbrough isn't somewhere you go to on a night out, is it?" When I suggest, "There's the speedway", Jamie gives short shrift "Late on a night!"

Keith McGhie is anxious about the heat 7 line-up, "Would somebody in the pits care to let us know that Robert Branford is coming out in blue?" Keen to help, Gareth Rogers advises – albeit still extremely muffled (as if down a well), "Peter Juul in blue coming out of pit gate." Sadly, Peter finishes last and the race effectively ends as a contest after Rusty Harrison easily overtakes Aaron Summers on the first corner. Post race we enjoy a brief burst of Kiss's 'Crazy Crazy Nights'. Keith segues his race results in with the music selections, "One man going crazy tonight is Rusty Harrison." Rat comes over all philosophical, "Is the interval after heat 8?" With his worst fears confirmed, Rat shakes his head disconsolately, "That's mad! It should be after heat 10 or 12. They've only just got here!" Keith McGhie reads a birthday announcement, "65 last Friday Bill Gimbeth. He's worked by Dave Rattenberry for many years. If you've been to the trackshop, he's probably served you! Wishing Bill A Happy Birthday From Dave." Nowadays Gary Havelock's in heat 8 and should regularly win. It certainly looks that way until Kenny Ingalls zooms past. Rat's simultaneously impressed and perplexed, "Who's that? Havvy shouldn't be getting beaten in these races round Redcar!"

Inside the hospitality area, sponsors and their friends enjoy the buffet, drinks, conviviality and the premium view available from the cordoned-off area of the clubhouse terrace reserved for them on race night, Steve Harland hasn't really any time to admire the racing. "It's going good – very busy! I could have written a book just about running the Redcar Juniors. More than most, I know how sensitive speedway promoters can be. I would say the people of Hull were fantastic to us. I know a lot of people didn't have a good word for Nigel Wordsworth but he was always great with us. Booey was as well. Rob Godfrey's approach was to annihilate everybody. If it was 5-1 in every heat, he'd be happy! Eric stopped them using Rob Hollingsworth which made things fairer. The whole idea was that people would see us and think 'Ah, Redcar Speedway!' Gareth Rogers did all the negotiations and CVS provided the money – otherwise we wouldn't have a club! Last year was very depressing in the Premier League, but we're still here and, as you say in your books, every year something more gets added."

With three wins from three rides, Comets reserve Kenny Ingalls clearly likes the STMP track. Tucked in behind the also impressive Richard Lawson, Kenny looks likely to further boost his pay packet with a paid win right up until the moment Jason Lyons – after spending a shrewd lap or so hanging back – blasts inside the young American on the last corner. It's a superbly timed manoeuvre that leaves no chance for recovery. Though he races for the opposition, Redcar fans appreciate what they see. "Until heat 9 – when he fell asleep – someone said to me 'their number 7 is riding like their number 1 and their number 1 is riding like their number 7!' I dunno what's happened to Doolan? They'd been having Ty Proctor but he's in Sweden tonight."

Sat by a table laden down with sandwiches, nibbles, pies and spring rolls for patrons of the sponsors' hospitality suite, long-time Bears fan Kay Ellis ("I was Morton in those days") looks more like someone's favourite aunt than a speedway fanatic. "I was Bernard Grant's – he was an announcer in the 60s, 70s and 80s – right-hand woman! I went on the back of Ole Olson's bike." Ian Drinkel jokes, "You traitor!" Ian helps out with hospitality work and also contributes to a column (co-written with Steve Harland) to the programme. Ian tells me, "If you want to know what went on behind the scenes and in the pits in the 60s Kay is the woman to speak to!" Kay's too much of a fan to sit down for too long. With great regularity, she dashes from the hospitality suite out onto the balcony to savour the action.

I'm enjoying my night at Redcar speedway. Steve Harland suggests I continue to ignore my critics. "I'd carry on writing! Better to be a bit of grit in the oyster. As a community protection officer, I get threatened every day. If a speedway promoter threatened me, I'd laugh it off." Steve enjoys some ongoing fundraising success via his commercial work on behalf of the club. "Take our advertising card – no it's not Lakeside (see I read your books) – and our fixture list card. I know it's a bit small. And we've got some posters, please take one. We've made a breakthrough after six years, we're into Middlesbrough Tourist Office! At their request!" [3]

Visitor Richard Lawson continues to impress. He zooms past the fast-riding Aaron Summers for a Comets heat advantage that reduces their deficit – after 11 races – to only 2 points. Marge Havelock tells me, "I'm making you another tea, actually I'm not, I'm going out to watch this race!" Marge dashes outside but isn't in time to notice Adam Roynon squeeze up Tomas Topinka or the Czech fall. Kay Ellis storms back angrily into the hospitality suite, "He was nowhere near him!" Marge asks, "What happened?" I'm too diplomatic to have an opinion. To widespread disbelief and ire from the Redcar faithful, Craig Ackroyd disqualifies Adam Roynon. It's not the best comeback meeting for Adam (with zero points). Once Kay calms down, she concedes, "Maybe he did drift a bit wide but he made a meal of it! Some riders do. We used to have one at Cleveland Park who fell off every week – Roger Wright! We used to call him Shirley Temple because he had such a perfect peaches and cream complexion. He'd fall off so he could be first in the bar at the Speedway Hotel. It was where we all congregated in those days. I've never seen anyone's complexion like it and he had curly hair. Bernard and I used to sit in the box where he had an old BBC-style microphone that you switched on and off. One night he left it on and I was shouting 'Fence him! Fence him!' not knowing it was on and we were told 'Could you ask the young lady to be quiet!'"

In the programme, each week the 'View From The Pits' column revisits a yesteryear clash between Middlesbrough and their visitors. Teesside Teessiders first met the Workington Comets in a British League Division 2 meeting at Cleveland Park in August 1970. In a page packed with tons of evocative recollection, information and some photos, we learn Brian Havelock rode for both Middlesbrough and Workington. Kay's particularly taken with the black-and-white team shot of the alliterative Teesside Teessiders. She can still remember all the riders. "Tom Leadbitter was nice and so was Dave Durham. My first full season was '64. It was the smell, I think – whatever they put in it – that got me hooked!" The rerun of heat 12 prompts further local disgruntlement when impressive Comets reserve Gary Irving and Tomas Topinka (second) combine for a maximum to give the Comets a two point lead (35-37). Steve Harland takes an oblique approach, "What do you think of speedway refs?"
[Jeff] "They do a hard job!"
[Steve] "What about Craig Ackroyd there?"

Before I can answer we're interrupted by an inaudible announcement about Tomas Topinka. "What did they say?"
[Steve] "They said his back protector saved him from serious injury. My wife's just said, 'They made him get up and walk back to the pits!'"

Fortunately for collective equanimity in this part of Middlesbrough, the Bears immediately hit back with their own maximum from Summers and Kus ahead of Kenny Ingalls and the comparatively hapless guest number 1 Kevin Doolan (who ends the night with paid 3 points from four rides). With two races to go, Redcar are 2 points to the good. Keith McGhie claims, "That could be a pivotal heat!" After a reprieve first time out because of a Rusty Harrison fall (unpopular with local fans), Kenny Ingalls beats Jason Lyons in the rerun, despite nearly coming unstuck on the last lap but somehow staying on. Incredulity and disbelief are widespread. "How did he stay on?" "Lyons gave him a 10-yard start." "They're gonna get a point from this meeting, aren't they?"

[3] Steve's programme column notes, "After some considerable amount of time we have finally made it into Redcar & Cleveland Borough Council's official tourist brochure – This Is Your Visitor Centre Guide. We could never understand why the council weren't including us in previous brochures because we actually lease the land on which the track was purpose-built on a motorsports site owned by them. Our inclusion will help us to build our profile in the area and beyond it. We've also made a breakthrough with both Redcar & Middlesbrough Tourist Office and Kirkleatham Museum is displaying a poster advertising us. We've also established a contact – Chris Colebrook, at Redcar, who has taken 250 brochures from us for visitors to take at their leisure – onwards and upwards."

If this were a Sky televised fixture, this last-heat decider would have Nigel Pearson and Kelvin Tatum verbally wet their trousers. Instead we just about hear a muffled Gareth Rogers apparently communicating with us from the dark side of the moon while being forced to wear a pair of oversized comedy false teeth. Keith McGhie quickly intervenes, "And the ref has a message: He's looking forward to starting heat 15 when somebody will tell him who the riders are!" We get to enjoy two races in one – Ingalls versus Kus and Lawson versus the wily Lyons. Viewed askance from the vantage point of the second-bend clubhouse balcony, it looks like a late burst from Richard Lawson snatches third for a draw. The riders definitely finish in a knot together so the referee requires the compound eyes of a fly to properly adjudicate. The fans split down party lines confidently predicting either Lawson or Lyons are third. After a brief delay, Keith McGhie relays the confirmed race result in slower than usual fashion "[theatrical pause] and third [additional pause] in red Jason Lyons!" Dave Rattenberry can't quite believe it, "That was never right but, then, we're at the wrong angle."

Back at my table Sunderland-based Redcar fan Brian comes up to ask, "Have you thought of writing a novel about speedway?"
"John Berry already has."
"Who?"
"John Berry – he used to be the Ipswich promoter and England manager."
"Never heard of him!"

Brian comes from Castletown. It's a mining village outside Sunderland. He used to sail for P&O out of Newcastle and also used to be a coal miner (he should write his own novel). Another Redcar fan from by the finish line says, "Lyons won by something approaching a bike length. I'd written it in my programme before they said. I think Adam Roynon has come back too early."
"He did well at Dudley on Tuesday before he fell off."
"He does that well! It's very different at that level. Look at Robert Branford, he does well there but struggles at this level."
"You were lucky they had Kevin Doolan."
"I don't know what's up with him. The first time I saw him here in 2009 he was very fast. Then the track has changed since then. It used to be billiard-table smooth but now it's rougher. It was very rough earlier in the season – I don't think that the riders liked it. Today they did! I prefer a dry and dusty track to a wet one because the racing is better!"

Whatever the exact moisture content, tonight's meeting is a great advert for the sport and kept the attention from first to the very last dash for the line. Out by the bus stop, quite a few fans wait to return to the city centre. For reasons best known to the bus company, at night the 64 and 64A buses transmogrify into buses numbered in the 700s. 12-year-old speedway obsessed Redcar (and Middlesbrough FC) fan Adam (and dad Stephen) are still on a high from what they've just seen. "That was an exciting meeting and what speedway should be all about! Last week was Plymouth – they were a one-man team. They were awful. What was good was the bikes from the 50s they had Japs and that. Adam saw the difference, didn't you?"
"Yeah, it was the noise – what a difference!"

Adam's an avid reader, particularly of sports biographies ("I like Michael Parkinson") and offers some thoughtful advice, "Why don't you do a book going to all the GP tracks? I think that would be very useful for anyone thinking of going!"

**28th July Redcar v. Workington (Premier League) 46-44**

# CHAPTER 17

## Scunthorpe

### "When the dirt builds up on that bend two and the riders go out wide, they can almost disappear in it up to their handlebars!"

29th July

Home pits

If you drive to see Scunthorpe Speedway, you meet friendly Scorpions parking operative Richard Heasman. Before he sought them out in Lincolnshire, he originally got his speedway kicks in East Sussex. "Is it true Alan Rolfe has left Eastbourne? I know Terry Smith well – he fixed my car before I came up here. Old Alan has been part of the furniture down there for years! I used to take the bikes out and, unlike some, I stayed for the end. Now I do the junior meetings here as flag marshal. It's a lot harder than it looks! Whenever I do it with someone for the first time, I always say don't touch that yellow and black flag until they've all gone past on the third lap – otherwise you pick it up and start to think 'now how many laps have they done?' Were you there that time the referee excluded Alan Rolfe? I've got this block and I keep losing it, I hope I've got it now – Frank Ebdon! I used to have that with Preben Eriksen. Frank used to like to run a quick meeting. We used to push the bikes so we saw what went on. They'd only just got back to the pits when the buzzer would go on to say they're on two minutes when he was there. One summer's day, he did Ipswich in 55 minutes! You weren't there years ago when he excluded all four second-half riders? And another time, as I say, he had Alan the start marshal off – I don't know what the problem was but he had Alan off the line!" Richard knows Eagles programme editor, Mike Corby well too. "He'd be embarrassed to hear but it would be good if everyone could speak to Mike! He's such a nice bloke. Whenever you speak to him, he gives you a lift!" Richard's work on the car park gates means he meets many people. "I walk round with Sally and lots of people I know from the speedway say 'hello' and stop for a chat." That's definitely the case with pretty well every new arrival. Car park charges are a talking point (even though it's only £1). One fan says, "It's £2 to park at Scunthorpe." "This is Scunthorpe." "I mean Sheffield!"

With 90 minutes to go, many fans arrive early to queue and then gain immediate access when the turnstiles open. Tonight the highly fancied and in-form Glasgow Tigers visit. So far the Tigers have won 15 of their 18 Premier League matches. In the pages of the *Speedway Star* prior to the meeting, Rob Godfrey shares his disbelief. "Surprised me? It shocked me! I don't even think they're a good team, although the stats clearly prove otherwise." With Richard Hall likely to return from injury and Scorpions reserve Thomas Jorgensen high scoring for fun, there would be grounds for confidence even against a full-strength Glasgow team. However, this Glasgow Tigers team are without all three heat leaders. They've lost Christian Henry through long-term injury (rider replacement operates), plus captain Joe Screen and a rejuvenated James Grieves miss out due to (sudden) illness. Some Scunthorpe fans still remain cautious, "We'll find out how good we are tonight, won't we?" Richard Heasman jokes, "It's an international match! We've got three chances – win, draw or lose." Richard appreciates the staff esprit de corps at the Eddie Wright Raceway, "Kenny Smith was the absolute tops when he was team manager! He always said 'hello' and 'thank you' for what you were doing and, of course, we also miss Rob Woffinden – though he didn't come here so much – but he is missed!" Joel Parsons' arrival delights Richard, "He's a nice lad! He's helping Richard Hall out now. He had a bad accident last year – he was very lucky!"

Over on the Scorpions side of the pits, riders and mechanics ready themselves in a calm effective way without fuss or frills. Much more at home in overalls than his BSPA jacket, Rob Godfrey stands by the tea urn with his wife (and fellow promoter) Gail as well as trackshop franchisee Dave Rattenberry. "Thank you for having me."
[Rob] "I didn't know you were coming, I've given up caring."
[Jeff] "I'll behave."
[Rob] "I'm sure you will, I don't mind what you do!"
[Rat] "The new stand is in at Redcar. They put it in today."
[Jeff] "It was there yesterday."
[Rat] "The seats weren't. They're really good – steep with a great view."
[Rob] "Isn't he going to charge £1 to sit there? Who'll pay that?"
[Rat] "I think they will!" [1]

Glasgow Tigers club presenter Michael Max stands and watches preparations in the pits. He bats off my greeting ("How are you?") with practised caution, "Surviving!"
[Jeff] "You must be pleased."
[Michael] "We're doing okay, for now! We're without three riders tonight, tomorrow and Sunday. James has bronchitis and Joe has an infection. We can't bring them back because people will think we're cheating. We're not [pause] this time! Why would we?"
[Jeff] "I thought this weekend was your 'weekend of truth!'"
[Michael] "It was, but the main thing is that Workington have lost twice away. That's massive!"
[Jeff] "Why?"
[Michael] "They're two wins behind us and they've six hard fixtures left."
[Jeff] "They were robbed last night at Redcar."
[Michael] "Even better!"
[Rat] "The Workington photographer claims to have a photo that shows it was Lawson third."
[Jeff]] "Really?"
[Rat] "Where were you last night for heat 15?"
[Jeff] "Same place as you."
[Rat] "Oh!"
[Jeff] "Mike Bennett is at Glasgow on Sunday."
[Michael] "I know."
[Jeff] "Last time he went elsewhere, they sacked the presenter."

[1] After spending Thursday night at the South Tees Motor Sports Park, it's something of a mystery why Dave Rattenberry goes back there again on a Friday morning. "Because it's impossible to get the stuff through the bar easily so, we stay locally, we can pick it up the next day and come here."

[Michael] "He's not coming in an official capacity, he's up to see his mum."
[Jeff] "I'm sorry I missed the previous Leicester presenter."
[Michael] "The new ones are good – a bit too chit-chatty – but they do the job."

Ex Sheffield promoter Malcolm Wright arrives early. "I haven't been here for a while. When Workington were here [looks in programme] in April. Last time I was here, there were only 400 here. I dunno why – the racing here is fantastic!" Barry 'Bazski' Preston rocks up just as Malcolm talks about last night's meeting at Redcar. "Dusty all night so you can't see, that can't be right, can it?"
[Baz] "The shale is better here."
[Malcolm] "He used to put shit down when he first started."
[Baz] "They got new stuff, the deeper they dig – the better it is!"
[Fan] "When he first started he was mixing that brick dust in, wasn't he, Barry?"
[Malcolm] "Doolan cost them the meeting really. He was shite. If they'd had a decent rider for James Wright they'd have won. Maybe he's too old?"
[Baz] "He's got a child and might be taking it careful? Mine's two months tomorrow – he went to his first speedway meeting at nine weeks old."
[Jeff] "What's his name?"
[Baz] "Charlie Robert."
[Jeff] "I thought you were gonna call the baby Rob after Rob Woffinden?"
[Baz] "And you got my girlfriend's name wrong in the book."
[Jeff] "What is it?"
[Baz] "Sarah and you called her Susie/Suzy."
[Malcolm] "I've never seen Gary Irving ride that fast, I was very impressed! I thought that Yank [Ingalls] looked brilliant! He didn't ride like that at Sheffield or the last time he was here – I think it was his first time. Lawson was good too. There's lots of funny things going on at Glasgow this season – allegedly – though not many riders would sign themselves off and miss three pay nights! It's Frampton they've got [for Screen], probably got him cheap because he's already up here."

Pat Hubbard works in the programme booth, while her husband Pete is a long-time member of the track staff. [2] Pat sells me a programme. She doesn't mention because Glasgow ride without their heat leaders, it is only three heats (2, 8 and 14) which remain unaltered in the programme but does ask about my well being, "How are you?" "It's a bit cold!" "You've been here before, haven't you?" Moments later – perhaps the definition of business optimism – the ice cream van arrives! Theo Pijper's wife queues for a programme with her child in a buggy, "They're just £1! They must be the cheapest in the League!"

Tonight's clash with Glasgow faces competition for local discretionary spend from a football friendly between Grimsby and Doncaster. Luckily for the Scunthorpe promotion, Glasgow travel mob-handed in their supporters' coach. Once inside the stadium grounds, Tigers fans congregate to quiz Alan Dick who initially meets and greets them by the programme booth. Alan is philosophical, "It's not good – three riders out: Christian Henry, James Grieves and Joe Screen." Soon the crowd builds, so Alan wanders with his audience to chat amiably close to the toilet block for nearly 30 minutes. In what other sport can you have such direct unmediated access?

After a 2010 season mostly notable for her rare appearances, the late Kenny Smith's partner Julie Harrowven once again relishes her job as marketing manager. "I only came four or five times last season. I didn't feel right but, as soon as I heard Rob had doubts, I had to come back! By the end of last season, it was alright! The town had been hit hard by the recession and the closure of the steelworks – things aren't good and that shows. We've done so many shows, fayres and

[2] Two weeks later the *Speedway Star* would report that Pete has been absent from the club of late through illness. Promoter Rob Godfrey sent him a Get Well message via the pages of the magazine. "My 'main man' Pete is not very well at the moment and is having a tough time, but he reads the *Speedway Star* every week and I just want to say 'Get Well Soon'. I need you back because the place is falling apart without you!"

exhibitions and mailed out tons of complimentary tickets. I wander round throughout the meeting and we have Terrace Talks so that the fans feel that they're listened too. It's a release valve. If you look at the car park, you can see. I get the comp tickets back from the girls and hear how many programmes we've sold. It looks okay tonight."
"There's a coach of Glasgow fans too."
"That'll be fun in the bar later! I think everyone still misses Kenny. There isn't anyone like him and he can't be replaced but we're working to get things going! I'm enjoying it and doing everything to let people know. Trouble is, sometimes the weather isn't good. We have a few weeks without a meeting or, like last weekend, a National League meeting."

David Howe wins heat 1 – sponsored by the Market Rasen Pet Centre – in a new track record time of 55.5 seconds. It's precisely the advert Rob Godfrey needs for the 1,500 tons of "new" shale that's available for sale at the club. The Eddie Wright Raceway develops and improves almost every time I visit. The latest visible addition is a protective mesh fence to form an additional protective barrier between the track and the fans. If there are to be overtakes then it appears these are only allowed on the third bend of the last lap. Michael Rajkowski bests Steve Worrall there in heat 2. Club announcer/presenter Shaun Leigh uses the time before heat 3 to big up my attendance. "Nomadic speedway author Jeff Scott is here. His latest book just like the rest of them is excellent! No doubt there's plenty of stuff in there on Scunthorpe. It's a very humorous and serious look at speedway from everyone's perspectives – very enlightening to say the least." Additionally, of course, there's a cornucopia of speedway merchandise in the trackshop. Copious stock is the hallmark of the club's franchisee, "In the trackshop Dave Rattenberry will have plenty of goodies too." Ever the salesman, Dave Rat immediately shouts from the doorway of the trackshop, "New Newport DVD at £8!"
[Jeff] "Fancy choosing that to shout about?"
[Bill Gimbeth] "It's the away win."
[Jeff] "What exciting stuff have you got really?"
[Bill] "Everything!"

By the start line a small knot of passionate Scorpions fans stand out since the ladies in their group (Marie and Michelle) scream passionate words of encouragement to the riders or share some precise but sharp advice with the referee. Richard Sweetman wins heat 3 pursued by Tero Aarnio who's serenaded throughout with words of encouragement ("Go on Aarnio! Go on Aarnio!"). Heat 4 sees Michael Rajkowski ride a very wide line for nearly three laps until he suddenly loses heart and concedes third to Steve Worrall – ahead, Richard Hall snatches first from Tigers guest Kozza Smith in their dash to the finish line. Lurking by the trackshop doorway, trackman Bob takes a slightly jaundiced view of speedway administration, "What's the point of a Rulebook? They tore it up years ago!"

Heat 5 sees Jordan Frampton get a flier until called back by Bradford-based referee Dave Dowling. The rerun sees Frampton twitch on gate 3 and, thereby, prompt Tero Aarnio to push into the tapes but fail to break them. Marie shouts "Frampton!!?!" and Dave Dowling illuminates the blue light to disqualify Tero. On the centre green, Shaun Leigh advocates the lessons of his childhood drama classes, "We'll try something new – musical statues at the start line until the tapes go up!" Marie wants these new but ad hoc advisory guidelines properly enforced, "Will you tell Frampton that! Cheat!" No sooner are the riders back under orders for the re-rerun than referee Dowling lets the tapes fly. Impressive Scorpions reserve Thomas Jorgensen appears to get exactly the flier that – when Frampton did this – referee Dave Dowling called "back". However, though consistency could be valued, the race continues. Marie and Michelle shout in unison to third-placed Scunthorpe Number 3 Michael Palm Toft, "Go on Toftie!" Despite the encouragement, Toft makes no impression. After the race, I ask Steve – the tall male member of said passionate group of Scunthorpe fans – if

the ladies are "members of the Toftie club?" Back in the trackshop, I check the evidence of my own eyes with Bill Gimbeth, "He got a total flier."
[Bill] "He did."
[Jeff] "A fraction longer and he'd have been through them."
[Bill] "There was no shouting then, was there?"

The third bend of heat 6 sees Kozza Smith cut inside Ben Wilson who falls. Initially Dave Dowling disqualifies race leader Nick Morris before changing his mind to exclude Kozza. Marie and Michelle helpfully shout, "That was Smith in case you didn't see it!" From heat 6 onwards – possibly to address twitchy riders sometimes obtaining advantage with a flier – Dave Dowling more or less lets the tapes up immediately. In the rerun of heat 6, Nick Morris makes a complete hash at the start, but recovers with exhilarating overtakes on the last two laps. Presenter Shaun Leigh salutes the skill of the Tigers reserve, "An absolutely fantastic ride it has to be said. Mr Howe and Mr Wilson had it all under control and young Nick Morris rode a great ride!" With itinerant Halifax versus Team Viking on in the second half, the man behind their possible rejuvenation in the town, Geoff Hayes, is at the Eddie Wright Raceway. "In the National Junior League, we finished bottom! Tonight this is the Midland Junior Challenge League and they're sitting pretty at the top!"
[Jeff] "Was that the one they didn't have permission for?"
[Geoff] "That's the one. The racing here is great!"
[Jeff] "I know."
[Geoff] "I must get over to the pits."
[Jeff] "I thought Wayne did it all?"
[Geoff] "We do it together. I supply the oil and fuel so I have to go and do that so they can warm the bikes up!"

Carefully picking his way back from the catering facility to his trackshop lair via a crowd bottleneck, Dave Rattenberry gets hit by shale as the riders roar into the first corner at the start of heat 7. With the rapidity of eye movement that sets apart predators from prey in the wild, Rat checks his chips for impromptu additional flavour. While he devours his mid-meeting snack with great gusto, Rat asks, "What have you written about me this time?"

Steve Worrall moves at the tapes at the start of heat 8 but, rather than see this as another signal that society has gone to hell in a hand basket, Marie and Michelle laud his skill, "He did well not to touch the tapes!" Sadly, Steve only gets as far as the third bend where he falls though his bike continues to somersault dramatically. Tall Steve has a practical solution, "They all move all the time. I think they should stretch some elastic behind them so that, if they move back, they get excluded as well as if they touch the tapes!" The rerun of heat 8 sees another win for Nick Morris, while Scunthorpe's power reserve Thomas Jorgensen wins heat 9 (11 points from four rides). Scorpions fan Steve recognises Jorgensen's third race win by clapping his programme board over his head like a seal (if they did that). "I think you've got to encourage them even though they don't probably see you!" Shaun Leigh luxuriates in the warm afterglow of another Jorgensen win, "Thomas Jorgensen is absolutely on fire tonight!" David Howe and Ben Wilson win heat 10 from the gate to put daylight between the Scorpions and Tigers at 32-38.

While we enjoy the track grade, Shaun Leigh issues instructions, "Would the second-half riders get to the pits please." Heat 11 starts dramatically when Jordan Frampton rears impressively and demolishes some second-bend kickboards. Dave Dowling waits to stop the race until the rest of the riders reach bend four. Without Frampton in the rerun, Thomas Jorgensen and Richard Hall hammer home another Scorpions maximum. Shaun Leigh notes, "I said it earlier, Jorgensen is absolutely flying this evening! And a bit of help from Rob there riding every inch of the track with him." Super reserves Jorgensen and Nick Morris recommence their personal duel until the Scorpion loses his steel shoe on the third bend of the third lap. Morris races throughout in thrilling on-the-edge fashion! So much so, he experiences various close encounters with the safety fence. With three races to go (after a drawn heat 12) the Scorpions enjoy a comfortable 8-point advantage. In the trackshop, Rat tries to persuade Ian Steadman about the virtues of my writing, "I'll say this, you can always pick up your book and read it, there's always something!"

Suddenly showing amazing turbo-charged speed David Howe easily wins heat 13. Scunthorpe gain another valuable point after Richard Hall remounts for the third place point available because Jordan Frampton retired. Before the penultimate heat, Sheffield's Eric Boocock gnomically observes, "Tactical coming up!" Rather surprisingly, Tigers team manager Stewart Dickson gives the black-and-white helmet colour to Richard Sweetman rather than in-form reserve Nick Morris. Keen to retain his ascendency from gate 1, Thomas Jorgensen seeks out and knocks into Nick Morris (though he's off gate 4). If he'd greater experience, Morris should fall theatrically but, instead, he tries to stay on and tumbles belatedly only to gain his own disqualification! Other times, even this might have been seen more sympathetically as meriting a rerun with all four back. Sweetman's third effectively ends any outside chance of a Tigers away match point tonight. Trackman Bob is quick to point out, "Sheffield will be smiling – Scunthorpe done them a favour! I was talking to those two people – they haven't been for two years – and I said, 'Aye, the rules are still stupid!'" Bob enjoys watching novice and expert riders alike, "The 125s at Sheffield were great – a bit scary, they're like milk bottles some of them!"

While we enjoy another track grade prior to heat 15, marketing manager Julie Harrowven passes, "This will put a smile on the faces of those that think Glasgow are cheating." Howe and Wilson easily triumph from the gate in a race where the only real zest takes place on the last corner where Tigers guest Kozza Smith aggressively overtakes his 'teammate' Sweetman for the proverbial chip chop money. Though level after nine races, the scoreline balloons to 54-36. Shaun Leigh strikes a philosophical rather than triumphant tone, "As we say every week, it takes two teams to make a match." Scorpions fan Steve enjoyed tonight's meeting so much that he nips into the trackshop to buy a copy of the Newport DVD, "I wonder if they'll have Richard Hollingsworth doing the track grading on there? We went to Newport a few years ago – nice stadium but I think that you have to get the racing right!"

The second half pits Halifax against Team Viking and immediately we get to see an exciting four-lap handlebar-to-handlebar battle between Adam Wrathall and Ricky Scarboro. Glasgow fans dribble away towards the stadium exit, "The ref spoilt it – he kept letting them jump all the time!" Neil Machin is sanguine, "Good win for speedway I'd say, mate!" Over six second-half heats, the four rider teams of Halifax and Team Viking remain evenly matched until a last race 5-1 for Halifax. For the cost of admission, fans see (excluding reruns) 21 heats in just under two and a half hours plus they get to enjoy fast harum-scarum racing and frequent overtakes. Each week Scorpions presenter Shaun Leigh relished what he saw of Glasgow Tigers promising reserve Nick Morris. "Tell you what, there were a couple of races where I think Morris were riding at the edge of his ability! That bike were too powerful too. When the dirt builds up on that bend two and the riders go out wide, they can almost disappear in it up to their handlebars! He found it and stayed on!"
[Jeff] "Is it me or does Jorgensen have the second biggest neck brace in speedway after Timo Lahti?"
[Shaun] "Me mum's dog had an operation once and came back with that on round its neck."

Shaun also works at Sheffield, Wolverhampton and Dudley. The recent Golden Hammer still sticks in his mind, "The crowd at Dudley was the biggest I've seen at Wolves for three years and I'm talking Elite League meetings! I was sitting in the box and Nigel Pearson came up and saw I hadn't filled in my programme and he said, 'Are you going to use that?' I said, 'I doubt it!' and he said, 'We could sell it!' I said 'it's only three quid!'"

**29th July Scunthorpe v. Glasgow (Premier League) 54-36**

# CHAPTER 18

# Workington

## "I'd take two broken legs over this!"

30th July

Studying the action

Three hours before the start time, it feels like the Costa del Workington at Derwent Park Stadium. Understandably there's little activity even in the cool shade of the pits. Glasgow Tigers Danish guest rider Claus Vissing's two bikes are unpacked, ready and gleam immaculately. Comets in-form American reserve Kenny Ingalls lurks somewhere else while his mechanic works, "He rode well at Redcar – he was pleased with it. In speedway, you're only as good as your last meeting."

Stood at the pits gate with clipboard in hand, Sean supervises all arrivals. It's my first visit since the late Ian Thomas's death. "It's still funny here without Ian everything is good but, still, something is missing! When he went to Belle Vue, you knew he'd be back. He was a friend anyway – we stayed in touch on Facebook – but this is different. When we had his book [Wheels and Deals] signing here I went along and I could see him scribbling and, when I got there, he gave me a book. He never gave anything away! When I looked inside it was to Delbow – that's my nickname – thanking me for all I'd done at Workington. I've been here since the start! I'm proud of my involvement with the Comets. My most treasured possession is a photograph with Lou Sansom and Ian Thomas and I'll be able to show my grandchildren and say 'Look I knew these heroes!' Ian was Ian. He was [taps head] clever. If you'd known him for a long time, you knew that he wasn't quite what he was with the falls and everything. The one thing he never did here was win the League and, this year, we have a good chance and they should dedicate it to him! I know he's looking down." Comets Clerk of the Course Rachel Robson interrupts Sean's reminiscences. She carries an authoritative slim attaché case and wears a fashionable pair of large lens sunglasses. Sean asks, "Do you know the teams?"
"Yes."
[Hands over the clipboard] "Can you write them out please? I was out at Rusty's and, though he shouldn't, he's gonna ride. He doesn't think everyone in team is as committed as they should be. He shouldn't but then he can have a week off. Who've they got?"
"Ostergaard and Vissing."
[Jeff] "Oh! Not Frampton and Kozza like last night?"

Rachel writes in injury returnee James Wright as Comets number 1. It's good news, though Workington still operate rider replacement for captain Tomas Topinka. Sean isn't a fan of this facility. "I don't agree with rider replacement. The fans should be able to see seven riders. Have we got a number 8?"
"Nah."
"You should always have a number 8 in case one of your reserves gets injured. Did you see Poole on Monday?"
"I didn't."
"It was the track not crashes but they didn't have anyone either. Instead of rider replacement, you should be able to choose any rider of the same average and use them as much as you want! So, if there's a 5-point rider riding at nine, use him – not the one with the five riding at only a 3-point average." Rachel leaves and Sean advises, "If I was you I'd go down by the programme booth – 30% of the people come in here [nods to main stand] but 70% down there."

The booth is adjacent to the pits. Casually drinking from an Evian Sports water bottle, Workington team manager Tony Jackson arrives before the majority of his riders. Tony's a lifelong Workington fan. Understandably team management is something of a dream job, albeit one with its own trials and tribulations. "I've had to make 24 calls today to get a guest. I've only four teams to choose from – Scunthorpe, Somerset, Redcar and Ipswich. I've got Tero Aarnio. I've got no Tomas because Roynon hit him and no Kildemand. The only winners today so far have been BT and Vodafone!"
"How did you see heat 15 at Redcar?"
"Well I saw it from the pits – which isn't the best place to see it from – and I thought Lawson got it. Fans at the line tell me that but you can't know? If it had been a yard further, we'd have won! Then we'd have won if Kevin Doolan had scored more than 2 points! Even if he has a bad night, you'd expect 6. He only got 4 at Plymouth for Ipswich last night. When he's not scoring, he's not earning."
"He's very professional. He even has that hydraulic stand thing."
"He didn't have it the other night, maybe that was it? Ironically with the rolling averages, he's still top of the Ipswich averages."

James Wright looks on while his mechanic Jonathan Bethell diligently attends to his back wheel.
[Jeff] "Lovely day."
[Jonathan] "It's fantastic."
[Jeff] "Costa del Workington."
[James] "A big cloud will come over in a minute."
[Jeff] "How's your mum?"
[James] "Same as ever."
[Jeff] "How's your knee?"
[Jonathan] "Well I went moto crossing to see how it and my shoulder was. I won my first two races and, people say, I fell off on my third. It was a night meeting the Wednesday before Cardiff and I don't remember anything until the next Monday! Nothing about hospital and that! I was talking to people and everything. I was out for 10 minutes but this is my first time back with James. I still don't feel right and my sleep isn't restful. I wake up more tired than before I went. I'd take two broken legs over this! It's awful. At least, it don't make the knee seem so bad. Glasgow haven't got Screen tonight, he's got a stomach upset and James Grieves is out too."

Once inside the turnstiles, fans can buy a programme or ticket for the 50/50 draw. Alternatively they can linger by the crash barriers that mark the fans' viewing area from the Glasgow side of the pits. Others head off to bagsy favoured positions on bends one and two or seek the shelter of the covered back-straight terraces. Paul Sermon works on the 50/50 draw, "We usually have a prize of around £150 which is £300 sold whereas Redcar have about £200 as a prize which just shows." Paul also sells coach tickets for Comets away meetings. "It's 770 quid for the coach to Leicester and we're selling tickets for the Newcastle meeting. Away from here I'm a car salesman at Ford. I finish at five, rush home and change out of me shirt and come straight down here. Every week I'm tired. I only get off one day a week." Paul breaks off to serve a lady with a complicated order, "She's reduced her coach order from four to three because she's fallen out with her boyfriend!"

Out on the track, SCB official Darren Hartley inspects the air and safety fences escorted by Clerk of the Course Rachel Robson. Glasgow fan Jim Fleming lingers on the edge of the Glasgow pits area, "What a time to get injuries, eh? It'll be even worse tomorrow [against Edinburgh]!" There are a substantial number of Glasgow fans here. One Tigers fan clutches a copy of the Times & Star, inside Keith Denham says he expects a win, "Some Glasgow riders haven't been for a few years and others rode unimpressively last time". Glasgow fans suggest their in-form reserve Nick Morris holds the key to the result. "When Nick made his debut at Berwick, they thought easy meat when the elbows came out but so did his! He might be slight but he can mix it!" Glamorous Paula and Julie run the programme booth with efficiency and bonhomie. Julie tells me, "It's been busy tonight. There's an agricultural show on in Cockermouth – which we thought might make a difference – but it hasn't."
[Jeff] "The Maryport Festival is on as well because I saw tents galore from the train."
[Julie] "Yea, that's on too."
[Jeff] "At least Glasgow brought two coaches."

Julie handles the casual sexism of a bloke in blue overalls with politeness, "You've got a nice tan, have you been sunbathing with no clothes on in your garden?"
[Julie] "No."
[Man] "If you had, I'd be round." Julie smiles but fails to answer.

Meetings start promptly at 7pm at Derwent Park. At 6.30pm, the bikes roar into life for 20 minutes of atmospheric revving, only interrupted by the tip-tip-tap of Ulrich Ostergaard's steel shoe as he enthusiastically jumps up and down for his pre-meeting callisthenic warm-up. In marked contrast, another Glasgow rider with ear defenders on stares off into the distance.

Locally based fan Richard Dalston with the Finnish girlfriend arrives full of the joys of summer. "When I went to Birmingham against Eastbourne I think I brought the Finns luck – Joonas was unchallenged throughout, even if he didn't make the start, and Timo got paid 7. Timo's got the talent and style – as you said – and was on the pace." The meeting also had some drama, "Nermark fenced Pedersen and his bike scattered across the track. I haven't seen as bad a one since Bo Brhel went over the fence years ago at Monmore!" The Golden Hammer meeting also impressed Richard, "I can't remember a crowd that big since '97 for a Wolves versus Cradley challenge or probably since the Ermolenko meeting."

Workington Comets (and Glasgow) presenter Michael Max betrays no sign of his Tigers allegiances throughout the night. After he gives newcomers a brief outline of some simple speedway rules, Michael continues, "The most important rule – you're here to have fun and enjoy yourselves and cheer on your team!" Richard Lawson wins heat 1 from the gate in a swift style that prompts Michael Max to exclaim, "That looks very swift and fast!" Announcer John Walsh confirms his time of 62.6 seconds "equals the fastest time of the season already held by Richard." Before heat 2, Glasgow reserve Nick Morris ostentatiously tightens his steel shoe. Once the tapes rise, Gary Irving tries to welcome him to Derwent Park with a massive elbow but, wise to this move, Morris backs off and – with nothing for resistance – Irving falls but quickly remounts (possibly to hide his embarrassment). Kenny Ingalls wins comfortably and it's another drawn heat. After a track grade, Ingalls is out again and fires away for yet another win, again by the proverbial mile. Tero Aarnio's arrival at the start line in place of the absent Peter Kildemand (ahead of a drawn heat 4) stirs Michael Max, "Peter only did one lap in Denmark on Wednesday and couldn't really hold unto the bike as he would like."

Comets programme editor, Barry Stephenson, drinks in the action from the rough ground that overlooks bends one and two. He's not a fan of the new-look Premier League second half of the season, "Mr Hoggart has changed his mind and says it's a stupid idea!" Workington still have numerous fixtures to schedule and complete. Barry quickly lists the opposition for the second half of the Comets campaign, "Plymouth, Ipswich, Rye House, Sheffield, Berwick and Newcastle plus Newport away in the Cup."
"Your cup overfloweth!"
"You could put it like that! I haven't been to Rye House since 1973. I just picked up that both Glasgow riders have been banned for seven days."
"Will that teach them a lesson?"
"I doubt it. If riders can choose not to ride on tracks they don't like, where will it end? It's mad! I don't know if we'll have a sport in five years, let alone ten! You should be okay at Glasgow tomorrow – even if it rains – because the Glasgow promotion are no longer allowed to make the call-off decisions."
"Who does it now?"
"Blind Jim McGregor, Willy Dishington or two other locally based referees."
"You could still win the Premier League."
"We won't win the League, the Tigers will! Things are different now. For example, since we were rained off at Leicester, they've signed three heat leaders. We would have won by 20 points against the side they had then. When Glasgow went there they won 62-28."

The possibility of a Glasgow win at Derwent Park recedes further after James Wright and Rusty Harrison ram home the Comets second maximum heat advantage in three races. Behind them, there's an exciting two-lap battle for third place between notional teammates Ulrich Ostergaard and Theo Pijper. Richard Dalston isn't surprised, "Ostergaard was rubbish at Birmingham. They took him out of heat 14 and put Vissing in." John Walsh lets us know James Wright won in 62.4 seconds, "We've just seen the fastest time of the season! It goes without saying that it's James Wright's best time here, his personal best!" Barry Stephenson exhales, "Oh, bugger. I'll have to change the programme again! I will because it's James. Before he crashed at Sheffield, he did the fastest time of the season at Sheffield." James Wright comes out again heat 6. While he trundles to the start line, I snatch a few words with his mechanic Jonathan Bethell. Just at that moment, James's bike misfires, "I hope I've got the carb adjuster right."
"Fast time for James in the last."
"Yeh and he weren't trying! He were disappointed when he found out."
"Why?"
"Because he could have broke the track record if he'd tried a bit harder."

Between races Workington promoter Keith Denham chats. "I told Ingalls after Redcar it's the best I've seen him ride since last April when he got 18 for the American Dream team at Newcastle. He broke his leg in August but now he's back to where he was before. Kevin Doolan was hopeless at Redcar. Tomas has a similar problem and he knows it. If you compare how he was two years ago, he's a totally different rider! He changed bikes last September and hasn't been the same since then." Heat 7 sees the Comets pair of Aarnio and Ingalls split by Dutchman Theo Pijper. Keith's got a lot of time for tonight's guest, "Tero's riding with his head, he's an intelligent rider. He wanted to come here three years ago." Performances on the track are one thing but costs and crowds off it are another. "We've got 800-820 here tonight. It's good that Glasgow brought two coaches – it makes a difference! We get big crowds for Glasgow, Edinburgh, Newcastle and Sheffield – anything south of there is hopeless!"

"You can't feel pleased with the second half Premier League draw."
"It's going to be hard to get people for those meetings."
"Rye House?"
"Not good."
"Plymouth?"
"Even worse."

Workington's geographical position doesn't help cost management. "We've had six rain-offs, so far. We get three forecasts and call it off at 8.30am because it's rural here. It's not Birmingham or Wolves – people come from a long way. We've only been wrong once." With the Comets already 10 points to the good, Keith looks on with quiet satisfaction, "We made the track different for these tonight. We made it much deeper so, no matter what team they brought, we knew they wouldn't like it!"

Tigers team manager Stewart Dickson endeavours to manufacture a fight-back via the partnership of Vissing (in black and white) and Morris. Though the Dane wins, experienced Derwent Park habitués Lawson and Harrison occupy the minor places to mitigate the damage. With his metaphorical Workington Comets cap jammed tightly onto his head, Michael Max tells us, "Claus Vissing is proving to be an inspirational guest and they'll be pleased because he's guesting for them tomorrow night." Glasgow's fight-back peters out in the next race when Lawson and Wright race the Comets third maximum heat advantage of the night. It's a remarkably easy win, something that doesn't escape Michael Max, "That was about as convincing as it comes – a little bit tough on the first bend and, after that, plain sailing."

Away from the speedway, Richard Dalston is a teacher. "Three o'clock on a Wednesday afternoon my school – Lion House – had an Under-18 hockey match against St Bees, when I hear the sound of a bike roaring in the car park and see it's Craig Cook practising his starts with a mechanic! How he got the bike up Cleator Moor High Street and into the car park I don't know!" Gary Irving makes a dashing start to heat 11 but comes unstuck on the fourth bend of lap two. He strikes the fence with such force that the loud bang is audible throughout the stadium. With trademark agility after any on-track accident, Michael Max trots swiftly to the scene, "That was horrible, horrible for Gary there!" Lifelong passionate Glasgow Tigers fan Alan Maclean takes a phlegmatic view of this evening's performance, "Screen would have scored more than Ostergaard. I've not really been conscious of him and Vissing has got more than Grieves would have." Sometimes Alan has the devil in him. "I posted on the Tigers forum last week to stir things up a bit that Sweetman will cost us the League. He lets the others through at the bend, which generally isn't a good sign in a speedway rider! Morris isn't frightened and Rajkowski sticks at it. Missing our heat leaders couldn't happen at a worse time with the Monarchs tomorrow [at Ashfield]. Joe Screen gets a four ride maximum every week but watches his shoulder. He'd ride in heat 15 if he has to but he hasn't had to, so far." After the crowd serenade a hobbling Gary Irving back to the pits, sole Comets representative Tero Aarnio wins the rerun of heat 11. Michael Max approves, "Well you never know what you're going to get with a guest. We've had our ups and downs over recent weeks and this is a definite up."

Loyal passionate Tigers fan – Jim Fleming – avoids the British round of the Speedway Grand Prix. "I don't go to Cardiff because I can't stand all those flippin' Union flags." I've an important philosophical question for Jim, "Do they have bins in this stadium?" Apparently not, "I asked the chap in the kiosk and he said they don't in case someone puts a bomb in them." Under the weather Rusty Harrison belies his injuries with another race win for 8 paid 11 from four rides. Behind him, Michal Rajkowski, Kenny Ingalls and Nick Morris deliver a four-lap thriller without actually changing race position throughout! Before heat 13 Gary Irving withdraws from the meeting, while prior to the race Michael Max sets the scene, "Claus Vissing has yet to taste defeat this evening but, off gate 4, he might just find it a little bit tougher!" As if gate position weren't already enough of an obstacle, his teammate Ulrich Ostergaard (in black and white) fishtails from the start to completely mess up his race partner. In heat 14, Rusty Harrison and Kenny Ingalls obliterate any vestigial arithmetic possibility that Glasgow might obtain a League point. Pragmatism and courtesy sees Comets team manager Tony Jackson offer guest Tero Aarnio the chance to boost his pay packet further with an outing in a purely academic heat 15 (he declines with a "problematic thumb"). Too late in the day Ulrich Ostergaard discovers his gating gloves to ride like a man possessed. Ignoring the defeat, Alan Maclean rushes to the pits to chat with his all-time idol – and father of Comets number 2 Richard – Steve Lawson. A lady Comets fan can't buy my book since she's already given away her surplus money, "An ex-rider who used to be here – Grant McDonald – has had a baby and that cost me £20!"

30th July Workington v. Glasgow (Premier League) 53-42

# CHAPTER 19

# Glasgow

## "He may be a smidgen over 17 but he rides like he's 35!"

31st July

Tapes up

Only an hour to go before the latest Tigers versus Monarchs Premier League local derby meeting, there's a subdued atmosphere at the Saracen Park pits gates entrance. Leaden skies don't help the mood, "The clouds are blowing this way. If there's rain over there, it's coming here!" Back from a lengthy sunny holiday in Florida, Alison Chalmers is also downbeat, "You're here to see our league campaign go down the drain!" Even at the best of times prior to the recent seven days' suspension of Glasgow's Joe Screen and James Grieves, Alison isn't overly respectful of the governing authorities behind British speedway. "I see the management committee has got it in for us! Rusty Harrison had an injury but he was riding yesterday, wasn't he?" At the programme stall, they stress value in response to price queries, "£2.50 that's all!" Not only are Glasgow fans keen to see conspiracy and possible defeat for their "weakened" team, but Mike Hunter of the Monarchs is also cautiously pessimistic, "In theory we should have, but our away form isn't good. I think Glasgow are becoming a team that are good at home and poor away!"

Referee Chris Durno travels widely because of domestic and international duties, often with his son Tom. "We went to Sweden on Tuesday. I couldn't do that when I was a contractor. It was brilliant! The contrast with British speedway is something. They had a crowd of 5,000 for example. There were nine or ten riders travelling over on the plane and back so the chat was good. I contacted the referee beforehand and, though we hadn't met him before, he let us stand in the box throughout even though it was a televised meeting. Afterwards the promoters thanked us for coming." Chris and Tom went to Derwent Park yesterday, "The meeting last night got better over the last few races but, until about heat 10, it was first from the gate. There weren't many there when you think Glasgow brought a hundred." Chris emphatically rejects claims that Joe Screen's ban is a form of discrimination. "If you miss an away meeting through illness, you have to miss your next home meeting so he had to sit out Scunthorpe and Workington too under that rule." Gordon Pairman's straight-talking "Third Leg" column in the programme reminds readers, "the speedway world is small and it is often sensible to think twice before speaking" then segues into comment upon

former Glasgow captain Shane Parker as a "good case in point". Chris explains, "He took lots of photos when they had a rain-off against Sheffield and posted them on his Facebook. Jim McGregor came down later and took some too."
"They can't make their own rain-off decisions now?"
"Nah, either the SCB official in charge or Jim McGregor or Willy Dishington decide. They got off quite lightly really."

Thirty minutes before the scheduled 4pm start time, Alison Chalmers can't quite believe the subdued atmosphere, "It's scarily quiet here! [pause] It is the end of the Glasgow Trades fortnight." If the crowd is below budget, at least the meeting is on. According to Ronnie Anderson, this isn't something that they can say on a regular basis over at Armadale, "It's rained in Edinburgh every Friday for the last four or five weeks and it affects the crowd!" Alison suggests he feast his eyes on the vista that is the glory of the Ashfield track. Ronnie guardedly acknowledges its splendour, "This track looks a lot better!" Soon off elsewhere, Ronnie misses Alison's words of praise, "He's brand new. He's pumped so much money into there. He really cares for Scottish speedway. He made the [daisy] wheel for us last season. Since he retired, he works a day or two at the [Armadale] track."

While we wait for the rider parade to begin, the announcer updates us on medical matters ("x-rays to his vertebrae and he'll be out for a minimum of two weeks"), suspensions ("unfortunately Joe Screen and James Grieves have had to withdraw from Friday, Saturday and today's meetings due to sickness") and admin ("I've been advised that a set of car keys have been handed in to the snack bar").

Monarchs team news that Adam McKinna will replace Tim Webster at number 6 reminds some of the giant literal and metaphorical hole left by Lee Dicken's absence. "It's a shame that Lee Dicken has officially retired – it comes to us all – he should have finished at the end of last season." The ongoing search for explanations about crowd size prompts one fanatical lady fan to remark, "Since Sweetman came in, the atmosphere has changed. You'd have thought coming into a better team would have lifted him but it hasn't! Still, you live in hope." Jim Fleming also isn't convinced either, "When they signed Sweetman, at the time I thought they could sign Mat Tresarrieu – he scored 17 when he was here for Ipswich. If Sweetman gates, he wins but it's not that often if last night [Workington] is anything to go by!"

Michael Max conducts the rider introductions with his customary efficiency and predicts the Tigers will get their "sternest test at home from the team who've been here the most times – three – this season, Edinburgh." With a passionate interest in rider injuries, Michael also needs no second invitation to provide a medical update, "We were told an hour ago that James Grieves would've been fit enough to ride today and he'll ride at Wolves tomorrow and will be back in his usual place next Friday against the Monarchs [at Armadale]."

After an abortive previous visit to Saracen Park, I finally see my first race on the new-look widened track. My initial impression is of speed and sufficient space for possible overtakes. Until he gets of out shape on the last lap, Josef Franc looks an inspired choice of guest for Joe Screen (and still finishes second behind teammate Theo Pijper). Though the crowd response is muted, Michael Max is quick to offer congratulations, "Well that was massive!" Heat 2 sees Glasgow repeat the medicine in a race won by Michal Rajkowski ("celebrating his last meeting at reserve before he moves up into the main body of the team") closely followed by the in-form Nick Morris. Although Michael Max talked the meeting up as a close contest on the rider parade only a few short minutes ago – with the scores less than poised (at 10-2) – he suddenly looks to the future, "When we race here against the Comets and Sheffield Tigers, these will be absolutely crucial fixtures!" Luckily for any neutrals and/or the small number of Monarchs fans here, Matthew Wethers and Andrew Tully reply with a maximum in heat 3 (with Richard Sweetman last). Heat 4 sees Nick Morris show strategic tactical nuance far in advance of his youth, something the home fans greet ecstatically. Michael Max also luxuriates in his talent, "He may be a smidgen over 17 but he rides like he's 35! He really did ride a wonderful and first and second bend." Understandably jubilant, Alan Maclean savours the action stood with his sons close to the start line, "You can really feel the power here before the start! I used to stand just in front of the tapes at Blantyre."

During a brief track grade delay, Michael Max grabs a few words with visiting fellow announcer Mike Bennett. Tigers fan since he went as a child in 1960, Mike's in the area to visit his mother. "It's always good to see Mike Bennett!" Speedway gods clearly don't quite agree that it's similarly good to always hear Bennett speak since Michael Max's microphone immediately malfunctions to the extent that it sounds like he broadcasts from the far side of the moon (rather than the far side of the stadium in the pits). All successful presenters must have a larger-than-life personality and a greater or lesser degree of self-obsession. Mike Bennett certainly has the latter in spades and loves nothing more than talking about himself, "In fact, it's a shame you can't see what I'm wearing on my feet – my Tigers socks is my tribute!" After some mercifully brief further chitchat Michael Max flourishes one of his trademark ("Have you stopped beating your wife yet?") style questions, "How is Elite League speedway compared to Premier League?" Surprisingly, Mike forgets his season-long delight at the superior quality of Elite League racing nowadays apparently served up weekly at the Norfolk Arena and, instead, is diplomatic and nuanced. "It's very, very different – good crowds, great racing but I still miss racing at this level. You can't beat a local derby to stuff the Monarchs – which is always a pleasure!" Interview over, Michael segues into a talk about a "speedway author extraordinaire" whose "offbeat" books are on sale here this afternoon ("You won't miss him, he's wearing the most ridiculous blue and white check shirt!")

Heat 5 ends with a Monarchs heat advantage (after a race win by Kevin Wölbert) but their brief fight-back effectively ends at the start of the next race after SCB referee Graham Flint disqualifies Cook for tapes touching. Announcer Jim Coyle MBE gets a broad ironic centre-green smile from Michael Max when he says, "Let's see what decision the Edinburgh team manager makes – will Craig Cook go off 15 metres or will they bring in Kyle Howarth?" With reserve Adam McKinna already in action, manager Alex Harkess's decision to send out Kyle Howarth appears to prophesy another Tigers 5-1 and, so, it proves. Howarth's valiant attempt to clear the track after a lap two third bend fall comes to nought since Graham Flint stops the race, disqualifies Howarth and awards the race.

Heat 7 sees Matthew Wethers in black and white off gate 4. When the tapes rise, Nick Morris (off gate 3) shows skill and virtuosity with his front wheel turned in the air to the right to pincer-cum-trap Wethers by the fence. It's a manoeuvre that sees Claus Vissing escape for the win. Since Morris finishes third ahead of Wethers, it also completely negates the Monarchs use of their tactical option. Jim Coyle MBE glories in the spectacle of what he calls "a thrilling heat 7" and, race results over, to rub further salt into metaphorical wounds Jim adds somewhat unnecessarily, "and just to confirm Matthew Wethers was on a tactical ride." Glasgow pile on further agony with another maximum heat advantage through Pijper and Morris to balloon their advantage to 16 points after eight races. Out on the centre green Michael Max grabs Josh Grajczonek for "the last time this season". Josh sounds deflated, "I went to hospital and didn't really get the news I wanted to hear. I thought I was doing really well but I'm not so I'm going to go back home." Prior to the season Josh aimed to reach the Under-21 World Championship final. Michael notes, "You had goals."
"I had goals! Hopefully, I'll have the pin out before I come back next year."
"I think at the start of the season it would be Elite League for you next season and, I thought, you'd go out in a blaze of glory with a Championship medal – which, of course, you'll still get!"
"It's another year in the Premier League next year and, maybe, then the Elite League."
"Does that mean we'll see you back in Ashfield next year?"
"I dunno yet, let's see."

Craig Cook gates sweetly at the start of heat 9 but then sees Michal Rajkowski squeeze past him through the proverbial non-existent outside gap. It's back straight manoeuvre soon emulated in kind by Richard Sweetman to loud acclaim from the Glasgow faithful. This harum-scarum first lap produces yet another Tigers maximum heat advantage (six in nine races). Michael Max celebrates, "And while they'd no answer to the outside line of Michal Rajkowski – wow the Raj!" Surprisingly Monarchs then hit back with their own maximum (second in ten races). Kevin Wölbert lines up at the tapes for heat 11 in the black-and-white helmet colour. Alison Chalmers is far from impressed, "They should have given Tully the tactical in the last heat not Wölbert in this." Alison screams as Tigers Morris and Vissing fire into the first bend placed first and second. Though Katajisto retires, Wölbert battles throughout – so much so Vissing just about manages to repel his last-bend cutback and dash for the line. The tiger mascot of the Tigers is so pumped up, he roars unintelligibly into the centre-green microphone. Alison reassures me, "that's Tiger speak!" Despite the 43-23 scoreline, competition remains intense. From gates 3 and 4, Tully and Rajkowski fire into the first bend with elbows drawn and locked together. Tully's speed, Kung Fu skill and balance see him emerge ahead and zoom off to win comfortably.

While the riders conduct their Speedway Riders' Benevolent Fund collection, some Tigers fans doubt the choice of personnel maximises possible donations, "There's too many Edinburgh riders there!" After the winning raffle ticket (£135) draw, Michael Max gets down to the serious business of a lengthy interview with team manager Stewart Dickson. "I know you've been itching to talk to us – there was a bit of doubt before the meeting."
"Yeh, definitely! I wasn't sure of the guests ... we love to win."
"I know people scoffed when you said that we had the best reserves in the League, if not ever, for quite a while."

After further back and forth, the vexed question of rider absences raises its controversial head. Usually well known for blunt centre-green opinions, Stewart errs on the side of caution, "They basically overruled what a doctor says!" Though Stewart doesn't go into specifics, he hints darkly of future protests, "They can think again if they think we're gonna lie down!" Back on the terraces, Alison Chalmers is similarly aggrieved, "Did you know James Grieves can ride at Wolverhampton in the Elite League tomorrow?" Stewart reveals collective rider anger is channelled positively, "It's just made them even more determined. I think we're okay here but I don't see another away win on the road the way the guys are going at the moment." Ian Maclean has mislaid his son ("I can't see Alan – the trouble is he knows hundreds of people.") but luxuriates in yet another stand-out performance from the Tigers young Australian reserve, "Nick Morris is quite something! The track's great – we couldn't have done this last season." With visitors Michael Max usually likes to channel his inner John Humphrys but, faced with team manager Stewart Dickson, he exhibits the ferocity Nicholas Witchell shows when faced with a favourite junior member of the Royal Family. "Are you confident we are going to win the league?" "Yeh, I certainly hope so – everyone in the stadium does – even the Edinburgh fans who'll want to 'cheer on' Scottish speedway over the last few weeks!"

Present controversial rider absences apart, Alison Chalmers believes the collective strength of the 2011 Tigers team provides their platform for success. "We can even win away when Joe Screen gets 6 points – that's the strength in depth when our number 1s not in double figures! He gets 12 at home but doesn't do heat 15. Did you know Richard Sweetman's average away from home is 2.8? I do feel sorry for Josh Grajczonek – that's what happened to Mark Loram – he had a broken femur. That's not good." Alison's equanimity quickly departs when Geoff Craythorne arrives. "I wondered where you were hiding! We've only got half a team because of the management committee."
"They don't do it to you more than anyone else, that's just Stewart. They don't do that, you don't see that."
"Birmingham are able to run rider replacement for Joe Screen on Friday night, yet he's only ridden

for them once."
"You guys always say that, yous do!"
"Never mind James Grieves is riding for Wolves on Monday. Why is that?"

If Geoff and Alison disagree about the impartiality of speedway authorities with Edinburgh's Alex Harkess in the BSPA chairmanship, they're equally at loggerheads over the track or the achievements of Glasgow speedway over the last decade.
"Jeff says fans at other clubs are pleased to see Glasgow do well."
"For 10 years, they've been very poor."
"No, in 2006 we were at the top and finished third".
"My impression is for 10 years you've had a poor team. Everyone goes on about the new track but it's very difficult to pass here. I agree it's fast but, when they get on the line, there's no passing. I think with the old track there were more opportunities. That's not good if you've only got one type of person who can pass – Screen, Grieves and Complin. There's still a hundred times more chances [at Armadale] than here. The only people who're saying the track's better and there's more passing are Glasgow fans because they're home orientated."
"I read your updates and there isn't much passing at Armadale."
"There is! Our track has been inconsistent so there's been lots of passing because of that. Look what we've won? In 10 years, we've only had one good team – in 2003. Okay, we won it in 2008 and 2010 but we were just lucky, according to people here. Glasgow fans think the world is against them. If you go to Workington, Edinburgh or Newcastle you don't hear the politics you hear here. You've got to remember that Glasgow fans are very self-centred and, after 10 years where we've had the Indian sign on them, it's been very difficult for you!"

Only three riders take to the track for heat 13 since the Monarchs Craig Cook is a no show. Alison Chalmers tells me, "He's gone, I saw him go to get changed before the collection!" With a disqualification and two third places, it's not Craig's best ever afternoon. Claus Vissing wins the race to complete a four race paid maximum as a guest for the Tigers. Matthew Wethers then wins the penultimate race of the afternoon as prelude to a battle royale for second in heat 15 with Nick Morris. It's a contest taken in good heart by both riders if judged by their enthusiastic post-race handshakes. Claus Vissing's heat 15 win gives him a paid five ride maximum and Morris's second cements his place as Tigers top scorer with paid 18 from seven rides. Over at the pits gate a gaggle of Tigers riders gather. Michael Max confirms the evidence of our eyes, "He's a big lad Claus but he's still getting the Glasgow bumps! It's a Glasgow tradition to give them." On the home-straight terrace, off-duty referee Jim McGregor strides off purposefully towards the car park clutching a smart attaché case notable for the ostentatious display of a SAS frequent flier badge.

After the rider parade, Michael Max struggles to elicit much conversational insight from soft-voiced reluctant interviewee Nick Morris. Pretty well all we glean is: "I knew Mattie was going good" Alan Dick is happier to chat and thank fans for the "major help" of a £500 donation, "Yes it's very important this donation from the Supporters' Club," Alan continues, "We come back well after a traumatic start to the weekend. We lost three riders within 30 minutes of each other. To lose one is bad, two is worse and three is a disaster." Michael then grabs "a few words with the star of the weekend" Claus Vissing who modestly bats off Michael's congratulation. "Of course, but I'm always delighted to take guest bookings." After sympathising about the lack of frequent races in the top tier of British speedway ("The Elite League can be a bit stop/start"), Michael showcases some more Janet and John style questions. "You like the track?" "Yes, of course!"

Ignoring the comfortable 58-32 win, as a true speedway fan Ian Maclean invariably appreciates exhilarating races and riders. "What about heat 15! Nick Morris beat Matthew Wethers and Andrew Tully who are both good riders, especially here. He's quite a talent! Sweetman kinda redeemed himself, but it's not good enough just being alright at home. Winning like that, it's been a long time coming so it was good. Did you notice that the number 1 gate wasn't very good?" Ian's son Alan networks his way around the stadium. "Michael [Max] said he was pleased I was here because James was upset that he wasn't allowed to ride and wondered what he could do about it? The question is what do the rules say? It's at the discretion of the management committee when it's a non-speedway injury but, Michael says, it's a formality that they always grant but didn't on this occasion. Michael didn't know if Edinburgh promoter Alex Harkess voted – he should have exempted himself – but even if he hadn't, Neil Machin had. It's obviously a legally flawed process but – given the lack of an independent body – in the real world, there's not much that you can do about it. My advice was to go out and score a maximum at Armadale on Friday!"

Alan Maclean is also a shrewd judge of speedway flesh, "Wethers isn't a very good rider but Tully is. It's no secret that Glasgow would like to sign Vissing. Obviously Nick Morris is very talented – and Grajczonek is never any bother – but I don't think he'll be back for Glasgow next season. They could loan him out to another Premier League team and sell him the year after to pay off the overdraft. Apparently they did think about getting Shane Parker to ride this afternoon but they decided against getting a Sheffield rider because of doing the shitty on them with the weather and forgetting he was staying up the night before." There's no problem with Ian Maclean's memory, "As he always did!"

**31st July Glasgow v. Edinburgh (Premier League) 58-32**

# CHAPTER 20

# Belle Vue

## "He replaces the other Pole who rode in Poland when he was injured!"

1st August

Box

Belle Vue Aces versus Coventry Bees in the Elite League in front of Sky cameras is a sideshow compared to the most important meeting of the week, if not the decade, for Aces management and fans. That public meeting takes place tomorrow night at Gorton Monastery to discuss the new stadium. Fans and locals alike are encouraged to attend to raise questions, express delight or air reservations. Belle Vue bosses David Gordon and Chris Morton make no secret of the meeting's importance. The trade press reveals this new stadium is to be home to Belle Vue Aces but also a centre of excellence to coach and nurture British speedway talent. Its designation as our National Stadium reflects the endorsement of the speedway authorities and underlines its potential future significance. In addition to news on developments and the opportunity to purchase a commemorative brick, fans will also (again) be able to meet guest of honour Ivan Mauger. He's flown in to re-promote his book and also launch the "Belle Vue 500 Club".

Inside Vue – the official race day magazine – carries a full-page advert with the punning strapline "A Vue of Our Future" and advises readers, "Be there. Be involved. Be part of it." Before this momentous meeting at the Monastery, there's still a speedway fixture to be run. Tonight's referee Chris Durno suggests that the absence of Emil Sayfutdinov short-changes Belle Vue management and speedway fans alike as well as Sky Sports and their viewers. "Sky will be at Belle Vue tonight for one reason and one reason only and that's to see Sayfutdinov yet he's not riding because he's in the Russian Pairs or something! When they signed him, Rosco made a fuss that he'd rarely miss meetings – only Swindon and when there was a GP. There was no mention of Russian Pairs! I'll ask him when I see him. Belle Vue have signed Jankowski for the exotically named Borys Miturski who never even rode for them. I don't have the paperwork on that so I don't even know that he will ride? I can't see Coventry winning tonight. Gareth Parry – who's seen a lot of racing at Belle Vue – tells me that last week the track was good but, for the three or four weeks previously, he's never seen it so bad and he's been going there for a very long time. I do feel sorry for Belle Vue because – when

I've seen the Russian at Birmingham and Coventry – he's pretty well doubled the crowds. Whereas they've not got him and Sky's here, so the crowd will be terrible!"

I also chose this meeting to watch Emil Sayfutdinov and, in addition to the double whammy of Sky's presence decimating the attendance levels, there's also the Ivan Mauger factor. He's back in the UK for yet another extensive book tour to again drink deeply from the udders of the fatted cow of British Speedway. Like me, off duty referee Chris Durno saw Workington race Glasgow on Saturday night and Glasgow race Edinburgh on Sunday afternoon. "I was even more disappointed with the Glasgow meeting than the Workington one! I'd really been looking forward to seeing the new track – and we stood up on the banking that gives a superb view of the track and the pits – and it looks like the new track forces everyone to ride the outside line. The inside of the first bend didn't have a wheel mark on it and there's a dip on bend three that causes everyone to shoot across to the fence. And, since the riders all take the same line, no one is going to overtake round the outside because they can't, as it's too dangerous! I've also never seen John Campbell or Alex Harkess take such a lacklustre approach to team management. They, like the riders, just didn't seem interested! I expected them to put a spoke in Glasgow's wheels but they didn't. Wethers and Tully got a 5-1 first time out and third time out but neither wore a black and white."

On the 204 bus from the city centre to Gorton, high black clouds threaten the humid sunshine. A damp car park and soaked Kirkmanshulme Lane track swelters humidly. With Sky cameras around, the chap on the gate is even more reluctant than usual to let in newcomers. Because of fire regulations, nowadays book displays are banned from the foyer just inside the stadium entrance. Instead, along with Ivan Mauger, I've been relocated upstairs to a balcony position with a panoramic view of the track. Sadly from a sound, smell and atmosphere point of view, it's behind the glass windows that run the length of the home-straight grandstand. Ivan Mauger's stall is already completely set up. In addition to his well-received autobiography, there's a range of memorabilia displayed upon his trademark black-and-white chequered tablecloth. There's a rather attractive pits crew shirt that – at a bargain price of £25 – I'm tempted to buy on nostalgia and style grounds before I remember that I'm actually not part of Ivan's team. Said Ivan Mauger pit shirt sports a label that says "Mauger Motorsports Uniform shirt" along with a prominent Jack Daniels logo. In a world of political correctness gone mad, advertisers make spirits look the most luscious drinks in the world and then – rather hypocritically – urge us to "drink responsibly". Ivan's wife Raye (along with another similarly aged lady) repeatedly adjusts and readjusts the books, leaflets and merchandise.

Alan Wilkinson's autobiography (From Two Wheels to Four) publisher Ian Corcoran lurks by my rather sparse book display table prior to his prearranged meeting with Ivan. Though aged 72, Ivan has the physique and demeanour of a much younger man (though his ownership of a small-sized pocket diary and spidery handwriting betray him). Ian's come along to firm up arrangements for Ivan's forthcoming visit to Alan Wilkinson's Barrow home. After double-checking his diary, Ivan Mauger tells Ian, "Saturday the 13th August we're at Workington staying at the Waverley Hotel. Sunday morning – 14th August – we'll go to Wilkie's. How long will it take?"
"It's an hour and a quarter to Barrow."
"And then we're going to Buxton, how long will that take?"
"Three hours."
"What's your mobile number? I have a UK mobile number." Ivan fiddles inside his tiny diary for said contact details. Meantime, Ivan hands over his smart looking business card. On the recto, it lists – in extremely small print otherwise they wouldn't all fit on – the key trophies and titles Ivan won during his long illustrious racing career.

All work by publisher Ian Corcoran and ghost writer Richard Frost into Alan Wilkinson's From Two Wheels to Four is gratis. Though Ian reveals his sales numbers, revenues, the unit manufacturing cost and monies (so far) donated to Alan, he says, "I'd prefer you didn't say exactly otherwise people won't feel so benevolent." Alan Wilkinson's extremely touched and grateful. "When I went back to their house they'd bought a new kitchen with the money. It's been a life-affirming thing to have done." Sports journalist, ghostwriter and Inside Vue editor Richard Frost appears from nowhere. "I was in semi-retirement until Ian persuaded me to do the Wilkie book. He's the driving force behind it. All I got was a lunch."
[Ian] "I think that was all I got."
[Jeff] "You've both enjoyed a life-affirming experience."
[Richard] "We did, absolutely, I enjoyed it for that!"

Resplendent in his Aces replica top, programme vendor John Thomas is drawn upstairs to see the goodies on sale on the Ivan Mauger stall as well as to glance at the six-time World Speedway Champion. After he leaves, Ian wonders, "What were his parents thinking calling him John Thomas?" Though the Wilkinson autobiography enjoys great success, Ian doesn't let the grass grow beneath him. He recently visited California to get "Bruce Penhall to sign a brick so we can auction it off for Alan. I also saw Billy Hamill when I was out there. The bricks have been very popular. Some people have bought £1,000 worth of bricks and others £500 worth!" [1] On his first visit to Alan Wilkinson's house, it was what wasn't there that struck Ian. "There was nothing in the house to do with speedway! I don't know if he wanted to distance himself from speedway but I don't think so!" Aside from his passionate ongoing promotion of the Wilkinson autobiography, Ian takes a close interest in speedway matters. ""'I agree with you about BSI – if you look at where they stage GP's, it all backwaters. I am going to Malilla but only because I have a friend there. My expectation is that they would be investing in ground breaking new venues and taking speedway to new territories such as Spain, India, Dubai in much the same way as the Moto GP has done. I don't see any of that investment happening. Surely that's where any profits from the likes of the Cardiff GPs should be heading – to promote the sport beyond the regular circuit of venues? I sometimes think that speedway is a lovely close family sport – a real community – but that also means that you can't express honest opinions or ask questions about why things are this way or that because everyone knows everyone else. As fans, we accept that riders change clubs each season so no one really says anything because they'll be elsewhere. But, looked at from the outside, it could seem to be too close!" Ian bats off my concern that Ivan Mauger will provide too much competition for my book display. "Well, that's one way of looking at it – the negative way! Why not see it as a positive because, if you base yourself close by to him, you'll get the footfall too?!"

Safely back downstairs at his programme stall, John Thomas is poised and ready to sell copies of tonight's magazine (£3) or back issues. Like many fans within the sport, John couldn't believe the goings on during the close season. "What was all that silliness with Coventry and Peterborough just because they didn't like the rules? Where would we be if everyone who didn't like the rules walked out? We wouldn't have speedway! What about when they took Ivan Mauger off us and gave him to Exeter? We had someone on an 11.5 average taken off us! Then we had Peter Collins and Chris Morton come through. The trouble is that there's nowhere for them to train anymore – so there's no one coming through now! Why do Poole have to sign all the riders who are from round

[1] On page 7 of Inside Vue advises fans about their "chance to buy a signed Penhall brick". "Do you want the chance to own a unique piece of speedway memorabilia? Double World Champion Bruce Penhall has signed a brick as part of Belle Vue's Buy a Brick campaign and has asked for it to be auctioned with the proceeds going to former Aces captain, Alan Wilkinson, and his wife, Jean. The signed brick is going up for sale on eBay and the package also includes a letter of authenticity, signed by Penhall, to say that it is genuinely his signature on the brick. Ian Corcoran, publisher of Wilkie's book From Two Wheels to Four, met the 1981-82 World champion on a recent trip to California. Says Ian: "Bruce was only too happy to sign the brick and the letter to help Alan. He is an old friend of Alan's from their racing days and contributed a very valuable chapter to the book, which has been a great success. Bruce was delighted to have the signed brick auctioned for Alan. It is something that should prove very attractive to fans with its connection to Bruce, to Wilkie and to the exciting new stadium project that is going on at Belle Vue."

here to down there? That said we have some from there up here – Jon Armstrong and Hopwood, who rode down there who won't make it."

"When do the turnstiles open?"

"Soon. I don't have my float from Jayne [Moss] yet. We all get our floats from her. She's not here yet. There's been an accident at the roundabout at Denton and I think she comes that way."

To my mind no Belle Vue meeting is complete without knowledgeable passionate speedway fan, Charles McKay. He rarely misses a meeting at Kirkmanshulme Lane (or Buxton) and also travels widely to speedway tracks elsewhere. The past weekend saw him watch the NL Fours at Loomer Road. "I've had five visits to Stoke this year and there's been two generator failures! The NL Fours weren't too bad since the delays were after the two semi-finals and before the final – while Dudley were arguing to change their riding order in between the semi-final and final, if you see what I mean. And they had another problem with the St John Ambulance. They had two kids with them and they had to go at 9.30pm. Have you seen the bloke on the centre green? Chris Simpson! He took to heat 16 to mention the sponsor's name! He thought Belle Vue still rode at Hyde Road. The only nicknames he knew were Belle Vue, Stoke and Dudley's and he didn't even know Hackney's! If you were representing the sport, would you wear a Stoke City shirt?" Before the meeting starts, Charles comments on various things including US speedway tracks ("Don't go to Costa Mesa, they weren't that good"), the Winter of Discontent ("I was at Nick Barber's auction in February I think it was and Allen Trump was saying all sorts not even knowing who I could be! 'xxx is a crook, yyy's corrupt and zzz are stooges'. It really wasn't good!"), *Bouquet of Shale* ("I read it on the way to LA"), Dudley ("They are Dudley fans not speedway fans!"), track conditions ("I went to the Under-19s at Rye House, as many people fell off there as at the Belle Vue/Dudley meeting but everyone complained about that!"), and the World Cup team finals. "Think about it, the four best teams in the world and all the best riders in the world bar two – that Russian and Hancock – so if you want to see the best that would be the one to watch but the Sky viewing figures were awful! There's going to be a cock-up with Sky. How are they going to get all the motor sports on on a weekend? There's Moto GP, Formula 1 and Speedway. The Sky viewing figures are down in 2011 for the Elite League. All that stuff on the forums about the BARB figures is silly!" With the meeting starting soon, Charles cuts short our conversation, "I must go down and stand on me box by the broken Coca-Cola machine."

Sadly, possibly because of its location indoors (even hotter than outside) or maybe every interested Belle Vue fan already bought his autobiography last season, prior to the meeting there's poor interest in Ivan Mauger's stall and merchandise. There's even less in mine. One of the few fans to linger is Irene Cocker. "Me dad came in 1928 when it opened and I started to come in 1945 – so I've been coming 66 years. I'm 67 now. I work here as well. I work at the Dogs on a Thursday and Saturday collecting the rubbish. I've watched Ivan Mauger since before he came to Belle Vue. He rode with Peter Craven. Speedway it's fantastic, it really is! You can't get this at football. Speedway is my life. I love speedway people, the atmosphere, the riders, the bar staff – they really are very, very good. I enjoy every minute of it! I had me son John – he died three years ago – in April '66 on the Thursday and I was here on the Saturday. Me daughter Lil was born on a Wednesday in '72 and I went to the speedway on the Saturday. I've been all over watching speedway: Holland, Poland, Sweden and Germany. Me husband died 20 years ago of a heart attack. He had dialysis at 41 and he had two strokes and went. If it was far away, I'd still go. I really do enjoy it! This new stadium will be brilliant! I can walk from Droyslden to Beswick. It's 15 or 16 miles and, four years ago, I walked from Blackpool to Manchester. Speedway – as I say – is my life!"

In a bizarrely delayed ceremony on the centre green, referee Craig Ackroyd presents Craig Cook with the referees' 2010 Rider of the Year award. The refs chose Craig Cook as their most exciting rider last season! Both Craigs pose for photos with the trophy before Craig Cook tells centre-green stadium presenter Natalie Quirk (stood behind her lectern and microphone) receiving this award is "very exciting". Though there's little reason to big up this run-of-the-mill Elite League meeting, Natalie claims it's a clash between "these two great rivals – Belle Vue and Coventry!" She also glosses over the absence of Bees number 1 Emil Sayfutdinov ("Due to racing commitments in Russia"). Sadly, some of Natalie's other comments are lost because of equipment difficulties, "It looks like the mic is having a bit of bother again!" I don't travel with my own box to stand on so, instead, lurk on the terraces next to Charles McKay. We overlook the run-in from the fourth bend to the start/finish line. Charles is bemused by the presentation ceremony, "It's a bit strange giving Craig Cook the Rider of the Year Award for 2010 in August 2011!"

Because of Emil Sayfutdinov's absence in heat 1, we're initially told Travis McGowan is rider replacement but – once we've collectively filled in our programmes – we're told it's Kenni Larsen instead. It makes no difference to the race – won by Chris Harris with Przemyslaw Pawlicki second and Craig Cook third. Though heat 2 is billed as Lukasz Jankowski's debut for Belle Vue, in fact, as Stuart (stood next to Charles) reminds us, "He rode three times for us before he got injured." Prior to the race, Coventry Bees fan Russell Lowry (father of Bees physiotherapist Jodie) joins us. "I said to David Rowe 'Is Rosco on drugs?' He can put in any of our top four riders and he chooses our worst [McGowan]; luckily he changed his mind!" Like the race beforehand, after a grunky first and second bend in heat 2 there's no real racing thereafter. Nick Morris wins. Russell is a big fan of the young Australian, "At 17 I've never seen a younger rider with such a brain! He sees what other riders are doing and takes advantage." Stuart accentuates the positives about Lukasz Jankowski's first race for Belle Vue in 2011, "The new Pole did alright for two laps." Charles prefers to highlight the ridiculous, "He replaces the other Pole who rode in Poland when he was injured!" Heat 3 ends as a contest for Ryan Fisher after he falls on the third bend. Charles McKay puffs out his cheeks, "It looks a bit greasy by the look of it." After the race ends, I can ask an important question, "Are the start line girls here just for Sky?" Slightly blasé, Charles replies, "No we have them every week."

Apart from her sports physiotherapy for the Bees, Russell's got much to be proud of about daughter Jodie, "She got a 2:1 in sports therapy at Coventry University. I think she wants to go and become a primary school teacher." If Ricky Kling were a Bees rider, he'd soon get treatment from Jodie after a heat 4 first-bend pirouette and fall. We enjoy a slight delay before referee Chris Durno's decision. Ken Wrench explains, "The ref is just looking at the replay on the Sky monitor." I'm sure that an experienced referee like Chris Durno would usually instantly disqualify the blue-helmeted Kling but – with technology at hand – professionalism (and diplomacy) dictate that Chris reality checks his initial intuitive decision. Though this fall looks harmless, Ricky Kling sufficiently damages three fingers enough to prevent holding his clutch and so withdraws from the meeting. While we wait for the rerun, Edward Kennett's slow reactions to the stoplights and red centre-green flags continues to bother Charles McKay, "It's useful being colour blind as a speedway rider!" Meantime, Russell learns the latest television news, "I've just had my brother text me saying he's seen Jodie on Sky in the pits." Rory Schlein wins the rerun though Kennett led first time out. Charles wonders if I've visited Beaumont Park. "If you go on a normal night, you can see but – if you go for the Fours – you won't be able to! I went to Edinburgh and, for the first four heats, there was no passing on bends three and four."

During the Sky commercial break, there's another track grade and the bonus, a Natalie Quirk centre-green interview with Ivan Mauger. "It's nice to be back at Belle Vue [frog in throat] and nice to see so many spectators."
[Natalie] "When did you get here?"
[Ivan] "Last Sunday."
[Natalie] "What have you planned?"

[Ivan] "We've got 12 or 13 tracks we didn't go to selling the books, signing the books and taking photos."
[Natalie] "How excited are you about the new National stadium?"
[Ivan] "I think it's going to happen. David [Gordon] sent me an aerial photo. It will be a multi-purpose stadium with basketball, volleyball, and a soccer pitch on the infield. I'd rather have a rugby pitch."
[Natalie] "We're told it is as well. It's going to be good, not just for Belle Vue but for the future of British speedway."
[Ivan] "We're going to have World Cup matches and British finals and, probably, eventually have a GP! Why not have two in England? I think it's very important to get behind the stadium."
[Natalie] "We do have a very important presentation – a special brick for Ivan presented by Chris Morton and David Gordon ahead of our very important meeting at Gorton Monastery tomorrow."
[Ken Wrench] "It was great to see Ivan there and, of course, Raye with him. We have a delay for the paramedic who's seeing Ricky Kling in the pits – he certainly took a very heavy fall."
[Natalie] "It's always nice to see Ivan back at Belle Vue."

Sadly Charles has no news about my complimentary brick, "You have to buy one! Have you won a world championship?"
"No!"
"Are you a member of the BSPA?"
"No!"
"Have you any dirty photos of Terry Russell?"
"No!"

Though we can't see the Sky pictures or hear the invariably stimulating insight and commentary from Nigel Pearson and Kelvin Tatum, mobile phone technology allows Russell Lowry to keep us fully in touch. "Ricky Kling is out of the meeting they're saying on Sky and Freddie Lindgren is at hospital with serious head injuries at Wolves." Unevenly rising tapes causes heat 6 to be rerun. Edward Kennett flies from the tapes pursued round the Kirkmanshulme Lane track pursued by Aces Harris and Cook until Bomber overdoes his chase and slides off on the second lap. Russell Lowry isn't surprised, "You could see that happening." Though Bomber remounts, Kennett, Cook and Morris are long gone. Natalie Quirk remains philosophical, "We haven't seen Chris Harris pop down here at Belle Vue that often. He's up against his old team Coventry where he used to be skipper." After he confirms the race results, Ken Wrench strikes an overly sombre tone, "We've some bad news [pause] Ricky Kling has withdrawn with a hand injury."

Coventry's Ryan Fisher is a wholly committed all-action rider whose throttle hand sometimes overwhelms his tactical nous. Comfortably third, for reasons best known to himself, Ryan rides heat 7 on the edge rather than take the steady-as-she-goes approach required to beat Aces reserve Jankowski. Instead he crashes. Russell Lowry is unsympathetic, "He's riding like an idiot and he saw Chris Harris do that." Seconds later Rory Schlein snatches victory at the flag from Kenni Larsen to turn a possible Bees 2-4 into an Aces 4-2. Natalie Quirk immediately hails Rory's triumph, "What a race! Rory Schlein providing one of the best races of the season, winning by a bike length." Though an Aces fan, Charles McKay is a stickler for the rules, "She shouldn't say that, it's the ref's decision!" Post race, Ryan's rush of blood continues to irk Russell, "Fair enough, Rory rode a good line and Kenni's only been here once but Ryan rode stupidly!" While we wait for more action, Russell suggests a future possible book title, "How about Come Rain or Shale?" With new club-by-club rain-off refund procedures in place for 2011, Charles McKay – where weather dictates – enjoys his own refund experiences. "Have I shown you my official Stoke letter about my refund for the Buxton meeting? It was handwritten on a scrap of paper. How do you think they paid?"
"Cash?"

"No."
"Cheque?"
"No."
"Doubloons?"
"No."
"Postal order?"
"Yes! Weird! It costs £1 to get one of those. It begs the question: Why don't they have a cheque book?"

The initial attempt to run heat 8 sees a lusty first-lap battle for third between Coventry teammates Przemyslaw Pawlicki and Travis McGowan that ends when an aggressive Pawlicki unseats himself. Russell believes it's mechanical gremlins rather than red mist that sees the Pole disqualified, "Something went on the bike – something came up, did you see it?" With the Aces Cook and Jankowski on a 5-1 prior to the incident, Charles is morose, "Under the unwritten rules of speedway, this will be a 3-3." Russell disagrees, "It won't because it's McGowan! Pawlicki is a bit like Andreas Jonsson – they're both either eternally sick or injured! A friend of mine – Matte Soderstrom – plays ice hockey for the Coventry Blaze. He's a nice bloke but he's got no teeth. The first time I met him, we got on to speedway. He said he went to school with Andreas Jonsson and said he was always a sickly child and never wanted to play ice hockey and that he had rich parents, who'd made their money in timber." Travis McGowan initially defies Russell's expectations until he concedes position to Jankowski on the back straight of lap two, "Absolutely pathetic!" This first Aces 5-1 of the night puts a metaphorical spring in the step of centre-green stadium presenter Natalie Quirk. She excitedly celebrates Jankowski's debut "second" though Russell's having none of it, "He's ridden here before!" Over the stadium speaker Ken Wrench has an urgent appeal, "Can we have an electrician to the pits?"

In the pits, Coventry are without their club physiotherapist Jodie Lowry because she's next to her father eating an unappetising cold quarter pizza with gusto. Since graduating from the trackshop into genuine medical behind-the-scenes involvement, Russell proudly spots glimpses of his daughter on Sky Sports. "The best one was last year's Grand Prix final when Bomber went back to the pits and all you could see was Jodie's backside. I had to watch it on Sky when I got home."

As a family the Lowrys offer encouragement throughout heat 9 to their talented Australian reserve in his battle with Charlie Gjedde. Russell calls, "Go on Nick, drop it on him!" while Jodie urges, "Go on Nick! Go on Nick!" Kennett and Morris both look quick. They race to a maximum heat advantage and level the scores (27-27). Ken Wrench again issues a polite request to errant camera users. "Can we stop using a flash camera on the fourth bend? It is dangerous and distracting and, if a red light comes on, riders will stop for that." It's an idea that amuses Charles, "Ha ha ha ha!"

I've not previously seen myself as part of the Coventry Bees shirt-buying demographic but, I have to say, they look quite stylish. "They're quite smart those new Bees tops – can you buy them in the trackshop?" Russell immediately disappoints me, "No, they're just for the team!" I don't anticipate Jodie would post anything negative about her beloved Bees on Twitter. Nevertheless, nowadays Jodie studiously watches her Ps and Qs since Big Brother – aka Mr K. Larsen – keeps a close eye on her account. "Kenni follows me on Twitter so I can't say anything bad about the team!" Russell Lowry isn't on Twitter (and has no formal position at Coventry) so has no need to err on the side of caution when it comes to criticism, particularly of the Belle Vue Aces. "Craig Cook 9 from four rides and Bomber 4!" I'm pretty new to Twitter but Jodie is a power user. "Did you see Freddie Lindgren on Twitter posing with his towel as low as possible and a hairdryer? How he dared – especially after his Swedish sausage advert (YouTube it, Freddie Lindgren and sausage advert) – I don't know? He's there smiling with a sausage!" Russell interjects "I was telling Jeff that Przemyslaw Pawlicki reminds me of Andreas Jonsson." "What, he's ridiculously tall?"

Though Jankowski finishes last in heat 11, Aces team manager Jim Lynch – in true Tammy Wynette fashion – stands by his man (though in reality, he has no choice since Kling, in Elvis-like fashion, has already left the building). Lynch is rewarded with a paid win for Lukasz Jankowski – second behind his teammate Charlie Gjedde after leading for three laps. Another last place for Ryan Fisher confirms that he's either out of his depth or frustratingly inconsistent. Russell avoids any further heartfelt criticism but, instead, barks occasional exclamations "Fisher!" as if chiding an out-of-favour recalcitrant pet. Charles McKay *faux* savours the performance of Belle Vue's latest Polish debutante, "I always said the Poles were great! I bet that Gjedde first is popular in Berwick."
[Stuart] "How many riders have we seen look good on their debut?"
[Charles] "Who was that who looked good on his debut and got 16?"
[Stuart] "Tobias Kroner."

With Harris, Schlein, Pawlicki and Kennett in heat 13, Jodie and Russell chorus in unison, "Bees in red, Bees in blue, Bees in white, Bees in yellow". Quite what this might mean is open to debate. This particular Bees versus Bees versus Bees versus Bees ends – predictably enough – in a draw. Four points in arrears, Aces "debutante" Jankowski lines up on gate 1 but has little impact (albeit slightly more than Patrick Hougaard) in a race won easily by the McGowan/Larsen partnership. This levels the scores and means – almost certainly to barely suppressed cries of ecstasy from Nigel Pearson and Kelvin Tatum in their television commentary booth – the live broadcast holy grail of a last-heat decider. Rory Schlein wins heat 15 in the fastest time of the night (60 seconds) and the meeting ends with honours even. Charles McKay isn't sure whether loyalty or tactical ineptitude determined Belle Vue selections in the nominated last race. "Can you explain why Harris was in the last heat?" Clutching at straws, I wonder, "Maybe it was contractual?" Whatever the reason, Charles remains disgusted, "I'd have put Cook in myself."

By 9.45pm most pit stalls are empty and with almost indecent haste the Sky commentary team have shot off back into their expense-account lifestyles as if leaving the scene of a crime. Nevertheless, the pits and surrounding area throngs with fans keen to chat with riders, staff or officials. Perma-smiled Alun 'Rosco' Rossiter poses patiently for photos as, nearby, does Chris Harris. Charlie Webster in speedway track inappropriate blue high-heels (also worn to the Goodwood Festival of Speed) stands next to her wheelie bag. She chats with all and sundry – signing autographs and posing for so many photos you'd be forgiven for thinking there'd been a photo call. She's affable, approachable and extremely personable. "I'm on Twitter. I don't do Facebook!"

Sky staff continue to break down the commentary pagoda and pack away impressive quantities of outside broadcast equipment. Just outside the pits gates, an impressive number of riders' vans are either packed or almost packed with bikes, toolboxes and the other essential speedway detritus of the itinerant life of international travelling riders. Charlie Webster eventually escapes from her adoring public and quickly retreats into the shadows and privacy of the car park. She makes an animated mobile telephone call next to an impressively massive Bentley (perhaps Rolls-Royce, I'm not so good on cars) with personalised plates. Perhaps, speedway makes some people wealthy? Or, more realistically, you have to be financially comfortable to have any long-term involvement in speedway promotion, sponsorship or management.

**1st August Belle Vue v. Coventry (Elite League) 45-45**

CHAPTER 21

# Lakeside

## "Is that your sixth times World Champion ring?"

12th August

White tyres

There are a huge number of empty tin cans bagged up ready for collection outside the Lakeside Speedway Office. So many, they attract the wasps. 'Scary' Sheila La Sage explains, "The can collection is for Marc Owen. He takes them to the recycling and that pays for his fuel home and away. I've got Bob announcing it each week and thanking them and telling the fans to keep them coming. It is good because it shows the support for the young star." Inside the speedway office, Bryn Williams one-finger types weather advice for would-be travellers reading the British Speedway Forum. General Manager Alan Sargent multitasks and copes with all the last-minute surprises his job inevitably entails. "Ivan Mauger was supposed to be coming next week but he rang up to say he's coming tonight."
"Did he come here last season?"
"Nah, it was a night when Briggo was here so he refused."
"Don't they get on then?"
"I don't think so."

Used to wintering in warmer climes "down under", Bryn's personal dress code makes no concessions to the warm British weather. "Wearing those shorts, you should be arrested! I've put it on the internet so maybe Parsnips will come!"

Due at Lakeside tonight are pressmen from the London media, Time Out magazine's travel writer and books' editor Chris Moss along with official photographer, Rob Greig for an ongoing series into London's overlooked cultural gems. Chris and Rob will see Swindon ride but also enjoy behind the scenes access to riders, officials and fans. Their official guide will be Lakeside's energetic and creative press officer Ian Smalley. I'll unofficially dog their footsteps. Sheila intends to be on her best behaviour for these media gentlemen, "They say Ian Smalley will look after you tonight."
[Bryn] "Are you Russian?"
"No, why?"
[Bryn] "Because he's got a party of Russians too!"

Inside the portacabin that serves as the nerve centre of Supporters' Club operations on race night, Sheila updates me. "Last week the Speedway Riders' Benevolent Fund collection was bad. We only got £109 and some of the buckets were empty. We're going to do it again this week because we have to have £200 otherwise the club have to put in the difference. It's not surprising because it was a bit tense last week here because one of the silencers had been tampered with. It all went on in the pits, Edward Kennett's dad was accusing everyone but it was only Edward's! I dunno why he needed to? I feel sorry for the other rider [Nick Morris] who borrowed his bike because he had his 3 points taken off him too."

Relatively adjacent to the Lakeside pits area, George and Linda Barclay set up their display table as well as attach information posters and action photographs to the fence advising on the joys of Speedway Training Schools – tutored by George – run at Lakeside. Next to their stall a gleaming speedway bike stands in a half-erected gazebo (part of tonight's SRBF fundraising). One of life's gentlemen and also a faith healer, George Barclay welcomes me warmly, "Hello Bob!" While he fiddles with fence ties, Linda tells me, "Nigel and Frances are setting up Mark Loram's bike and the fans can sit on it for 50p or the big kids for £1! They're doing the Ben Fund again tonight because it didn't really didn't happen last week because of the weather and with sealing the pits. Now Ivan Mauger's decided to turn up this week when it was supposed to be next so people won't have their money or, if they do, they won't give as much to the Ben Fund as they might have otherwise! Did you see Davy Watt put a photo of George in an old monkey mask and put it up claiming it was him!? They found his picture in a Services and they were just having a laugh. I think it was a Reader's Digest article they did a good few years ago. I always liked Davy from his Rye House days." Gusts of wind threaten the gazebo so George calls for an extra pair of hands, "Bob! Bob!"
[Linda] "It's not Bob Coles, it's Jeff."
[George] "He's better looking than Bob Coles."
Linda thinks they should move Mark Loram's bike to firmer ground, "If the kids are going to sit on it, it's best it's on the flat, isn't it?"
[Nigel] "We have the gazebo but the wind is coming from the west."
[Linda] "It hasn't been as cold these last few weeks."
[Sheila] "All I know is that if we're winning by loads, the riders will go round with the bucket."
[Linda] "Alan reckons there's a chance of rain tonight at an hour in, so they won't want an interval."

After the weather, talk turns to rider injuries. Linda updates Scary on Stuart Robson's condition. "I know Rob had to go back in though and have some blood drained from his lung. He's home now and getting better."
"What about Adam?"
"He's not in hospital, he's at home too. He wants to come back before the end of the season and say his goodbyes to everyone."
Resplendent in one of the less stylish Lakeside Hammers anoraks, programme contributor Sid Shine joins us. "I do tables that no one else does. I do about ten on the website and three in the programme. You can see them on page 16. I do different ones to the ones you can get elsewhere. I love it here, I really do! It's a happy-go-lucky place and everyone talks. It's only 30 minutes to here and it was two hours to Eastbourne. I did ten years there but, when they changed to a Thursday, I couldn't do it any longer anyway. Cam and Watty – the two riders I liked for Eastbourne – are riding for us here tonight too! Their number 8 might be Sedgmen. His meeting up North has been cancelled so they might put him in or Tomicek but we don't know yet? There's a big group of Russians coming tonight."
"Where are they from?"
"Russia! I'd have thought they'd have come last week when Sayfutdinov was here."

While I wait by the entrance gate for Time Out, a steady stream of riders and staff pass in their vehicles. They sign in or show their pass. When two blokes walk up (one with a satchel-cum-man-bag and the other with a camera), I guess they are the media visitors. They travelled to Lakeside speedway on public transport ("I've a card I can use to get here") and then walked 20 minutes from the evocative Chafford Hundred Railway Station (serviced by C2C mainline) via the Lakeside shopping centre car parks and various arterial roads. Most importantly, they survive the perils of a journey on foot and want to savour the delights of a night at the speedway.

News that Ivan Mauger is here and, possibly, available for a brief chat, prompts Chris – seasoned travel writer and reporter – to pluck a spiral bound notebook from his bag. Already interviewed too many times in his long career to even start to possibly count, Ivan turns away from the neatly presented merchandise on the display table adjacent to his van to answer the latest series of penetrating media questions. Similarly well practised, Chris opens with a gentle loosener, "How do you feel about modern speedway, Ivor?" I'm fairly certain Ivan's going to tediously recount how things have changed since his day. Strangely, though he's thousands of miles from home, Ivan's journalistic antennae is so sophisticated he immediately recalibrates his conversation to the geographic location of the interview. Professionalism is the watchword of Ivan's career so this shouldn't surprise but nonetheless does. Ivan effortlessly segues into a thumbnail sketch of speedway in London in the '70s. "In London there were four or five tracks until the '70s – until 1974 – and this is the last one operating in Greater London." Ivan's description is so apposite that – when the article appears in Time Out (issue dated 8-14th September) – the decline in the number of London tracks is one of the key narrative optics Chris Moss uses to describes his Lakeside visit. Chris listens to Ivan with interest and questions incisively, "Does that make you nostalgic?" Ivan needs no second invitation to launch into detailed description of subjects close to his heart: book tours and money (in this instance, rider earnings). Rather than just a quick journey round 'some of the great pay packets I have known', Ivan also throws out some big numbers. "There were 272,000 people coming to speedway every week in the '70s. We used to get 90,000 at Wembley and at Poland it used to get 100,000 and sometimes 20,000 people would be out on the road listening when we were over there. It really was the biggest sport!"
[Chris] "How are you finding it nowadays?"
"It's tricky because the tracks mainly run on Thursdays and Fridays and Saturdays."
Rob Greig joins in and asks, "Is that your sixth times World Champion ring?"
"It's nine times."
In miniscule type, Ivan's business card lists some of the silverware highlights of his career. Moving on quickly, Rob suggests he take Ivan out onto the track to pose for photos. Notebook in hand, Chris turns his attention to Raye Mauger and quickly learns, "Ivan set up the table and I'm looking after it and I have the last word – like I do in the book!"
"How long are you here?"
"We're here until August 23rd – six weeks."

While Rob completes his photographic session, Chris and I linger by the stairs to the referee's cum announcer's box that overlooks the start/finish line. With thousands of interviews already under his belt with the great, the good, the dull and the culturally significant, Chris avoids some interviewees. "I don't want to meet David Bowie because he's an idol. Bryan Ferry is another. Allegedly, he's a plonker." Rob returns with a selection of photos that, ultimately, don't make the final cut into the magazine article. "I was just saying on the way here that I've never been to speedway before and the only riders I know or have heard of are Barry Briggs, Ivan Mauger and Peter Collins. And, on my first ever visit to speedway, the first person I meet is Ivan Mauger!" Arguably Rob should previously have been to speedway, "I come from ten miles from Swindon. I've been to Blunsdon to see the dogs but not the speedway!" Keen to meet characters and learn colourful opinions, Chris Moss bustles up the steep stairs to the announcer's box where he finds soundman Steve Gobey. I wait downstairs with Rob. He left college in Bournemouth 20 years ago and has been at Time Out for the last decade. "The year I went [to college] it had just opened and all the equipment was amazing and the next year you couldn't get on it. It was next to a ramshackle building called the Dorset Institute of Higher Education which is now the University!"

The next interviewee Chris finds is Scary Sheila La Sage in the Supporters' portacabin. He's an affable and voracious questioner. Chris establishes eye contact with his prey while he scratches out efficient shorthand on his pad like a bird pecking for worms after a heavy shower. Despite low key and affable questions, the normally uber-bubbly and effusive Scary comes over all shy. Scary's American husband-to-be also provides his impressions of the appeal of speedway at Lakeside. We then head to the pits area past the grandstand arcade of shops, Mark Loram's bike, George and Linda's Training School display table and the second-bend refreshment kiosk. Chris returned this morning from Japan after reporting on the aftermath and recovery from the tsunami. He betrays no sign of jetlag. "The North of Japan is wild and beautiful and isn't going to be rebuilt – unless tourists go there! They get people going to Tokyo and Kyoto but not the North. They're very good at rebuilding things but they'll need tourists to get back on their feet."

The home and away sections are crowded with riders, mechanics and helpers in the pits. Mostly the mechanics work unsupervised. Scott Nicholls chats animatedly in the far corner of the Swindon section waving his flat-screen Twitter-enabled phone around for additional emphasis. Since the riders are still in their casual clothes, it's hard to tell them apart from anyone else. Some riders stand, others sit silently unavailable or chat animatedly with colleagues and rivals. Rather politely Rob asks, "Can I get photos now or later?" Since speedway is a dangerous environment, Clerk of the Course Fred Parkin responsibly insists on insurance formalities, "Can you sign the disclaimer?" before giving advice and instruction. "If you go out onto the centre green, you have to stand inside the white tyres. Not literally. Last year for the Under-21s I told the start girls that, and they were trying to get inside of one of the tyres! Can you believe it?"

Smartly attired in his stylish Lakeside anorak, hardworking Lakeside marketing manager Ian Smalley arrives to escort Chris and Rob. "I was in the bar chatting to a major local club about a potential tie-up." Away from the sport, Ian is a creative director for an advertising agency that deals with big brands and major household names. "It's the age-old thing, I asked myself in 2007, 'how can I get involved in such a fast sport?'" With marketable skills aplenty, Ian contacted Stuart Douglas and Jon Cook to offer his services at the start of their Lakeside 'revolution'. The chemistry works plus the off-track marketing, publicity and commercial results have been impressive. Ian deserves his place on Lakeside's roster of club officials. Ian's people skills come easily to him. His knowledge of the sport, Lakeside riders, officials and staff make him the ideal liaison person to ensure media and commercial visitors enjoy the club's hospitality. Ian suggests Chris should interview club captain Lee Richardson but, with pre-meeting preparations in full swing, this isn't immediately a goer. Sensitive to both media and riders' needs, Ian's a model of diplomacy and helpfulness. "Lee isn't free now but he should be at the end of the meeting depending on how it has gone."
"What time does it finish?"
"At 9.30pm. Maybe we can grab Rob Mear, he's our reserve, and from Hitchin."

Time Out magazine's target audience is primarily Greater London based so, where possible, Chris prefers the local angle. He takes copious notes during his interview with club co-promoter Jon Cook. Afterwards, he says he learnt about "the business side of things, his ambitions for the club and what they're trying to achieve. It's nice to talk to someone so positive who can speak so crisply and concisely about their work and the sport!"

[Ian] "We can go out on the centre green about heat 4 so that you can see what that is like."
[Jeff] "I can explain the programme to you later."
[Chris] "I don't think there'll be any need for that!"

Not yet in their kevlars, the Lakeside team pass en masse chatting animatedly as they return from their track walk. [Chris] "They all look very young to me! Early to mid 20s, some early 30s."
[Jeff] "Some of them are old by speedway rider standards."
[Rob] "What young for the top level? Do they ride at 18 in the Elite League?"
[Ian] "Some can be 16 or 17."

Journalism wouldn't be journalism without a trip to the bar. Lakeside is blessed with a grandstand-cum-clubhouse bar balcony offering a panoramic view. So far, the access to and calibre of the interviewees pleases Chris. "I've got my story now. It's been lovely to chat to so many positive people. Everyone is an optimist!" Beers in hand on a sunny August evening, Rob comes over all wistful. "I just got the most fantastic shot of two old blokes. It was brilliant! If you wanted to pander to people's misconceptions – which I, we don't! – then that would be the shot." Also out on the balcony various small knots of Lakeside fans – some in replica tops – chatter but quickly clock the tall stranger with the notebook and satchel and his friend with the bagful of photographic equipment. Speedway fans are equal opportunities meeter and greeters so, within a minute of their introductions, Chris and Rob are subject to a lava flow of comment, insight and explanation. With London accents galore, Chris is spoilt for interesting characters and possible local colour. Just in case, he asks if any of the fans present – in addition to their proclivity for speedway – also indulge in any other hobbies, passions and pastimes of yesteryear? "Do you keep pigeons?" It fails as an opening gambit, "Nah, my granddad did as a kid!" Fortunately after a few more gentle probes, the conversation completely avoids any talk of Pearly Kings and Queens, wrestling, bingo or jellied eels (possibly on sale at the shellfish stall downstairs). These Lakeside fans of a certain age prefer to launch into animated discussion about great London speedway clubs of the past (that said, many statements are so garbled and/or inaccurate that I don't envy Chris trying to make sense of his notes afterwards).

Chris has many questions, some without answers, "When did West Ham speedway shut?" elicits a quizzical look and a shrug, "Dunno, a while back, he'll know!" He doesn't but, instead, goes long on the glories of London speedway clubs in days gone by. With so many men keen to provide insight for the journalists, things soon take on a Vision-On-meets-Marcel-Marceau flavour. One generously sized fan happily mimes what a speedway riders do on a bike when firing into the bend at high speed. "You get into the corner like this and, with my back wheel, I can take your front wheel out! The Poles and the Russians used to do that deliberately. They can't do that nowadays because it's not allowed." Though we live in a much more visual culture – the arcana of speedway minutiae, habits, rules and regulations would defeat the mime skills of star pupils from RADA. With the meeting yet to get underway, another fan helpfully launches into a lengthy almost mind-numbing explanation of gate positions, scoring, number of rides allowed, nominated heats, the use of guests and the rider replacement facility without once grasping at an imaginary throttle or an outwards thrust of a buttock in full broadside.

Prior to his off-the-cuff rider introductions over by the start/finish line, Jon Cook embraces his theatrical alter ego to bellow into the microphone, "Welcome to the home of the Hammers!" Rob leans over to Chris, "Write this down, they're playing the Ramones! Earlier they were playing 'London Calling!'" Scribbling out yet more spidery shorthand into his notepad, Chris double takes, "Is that Jon!?!" Transformed from his rational softly spoken pits persona, our master of ceremonies continues, "We're in the presence of greatness, the greatest rider of all time – Ivan Mauger!" After some more superlatives along with flimflam about Ivan's latest British speedway book tour, we get

to enjoy some more insights from a true speedway legend. "Yes, I'm enjoying it so far. We've sold a lot of books tonight. We're over there by the van. I've sold a lot before the meeting even started!"
[Jon] "Had you been riding with the GP Series – I know modesty will stop you answering – but I'm sure you'd be in double figures."
[Ivan] "I'd love to ride in the GP Series!"

Sadly the need to stage a speedway meeting soon curtails Jon's conversation with Ivan. By renown, Jon's rider introductions are nuanced but positive, effortlessly mixing insight and in-jokes ("He's flown in tonight from MON-ARK-CO, how cool is that? – Cameron Woodward!")

From the razzamatazz, strangers to the sport like Chris and Rob might think this meeting with Swindon is the most exciting night of the year on the Lakeside speedway calendar. Out on the balcony talk of defunct London speedway clubs slowly transmogrifies into a low key lament about changing societal values (a discussion prompted by the recent rioting in London). Amongst the geezers, the consensus is that community-minded men wouldn't allow such a situation to arise let alone to spiral out of control, "they would have sorted out the rioters in the '70s without the use of the police". For reasons I can't quite fathom (since I join the conversation half way through), Rob Greig suggests to one fan, "You shouldn't be reading the Mail" "Why not, I read the Mail and the Sun." Talk of exotic reading materials prompts one fan to whisper to me, "What's this magazine then?"
"Time Out, it covers the arts, books, culture and media as well as providing listings of what's on in London and where to eat, drink and dance."
"Where can I buy it?"
"At your local newsagents."
"I live in Essex."
"Smith's then."

From our panoramic view of the speedway track, we see the Elite League meeting fire positively into life (if you're a Lakeside fan) with maximum heat advantages in both heats 2 and 4. Hammers 5-1 signs flutter in celebration. At a table inside the bar with easy access to the balcony doorway, Bryn Williams types frenetic text messages into his mobile for the Speedway Updates site. Bryn's not been to so many meetings this season, "I've cut back this year – I've only gone to 98 meetings."

Their opportunity to watch from the centre green soon lures Chris and Rob back down to the pit lane. They chat to Ian Smalley until they can safely nip out to the centre green. It's as close as you can get without actually racing a bike for a living. Unlike old-fashioned or hypersensitive speedway promoters, Lakeside are *au fait* with print, broadcast, and online media opportunities. They have a top-notch award-winning official programme; Stuart Douglas and Jon Cook always give good copy; the rebrand of the club adds lustre and works well across all their communications while, under Ian's auspices, you can download a free e-zine (Hammerzine). Ian tells us, the social media du jour – Twitter – is an opportunity already seized, "I did an article covering Lee Richardson's speedway tweets for a month in one of the back issues of Hammerzine. It's amazing how much he tweets." Ian believes that some clubs miss out on obvious publicity opportunities. "For such a fast sport, some promoters don't appreciate the speed and power of the internet. Equally they don't realise the danger. When you've got something like that, you've got to harness it for your benefit! You can't shut the internet if it doesn't always say what you like." If speedway excites from behind the safety fence, it thrills on a different level from the centre green. No matter how many times I enjoy the privilege, it mesmerises me.

Understandably, Rob and Chris express delight in the spectacle from this perspective. Riders power into bends in full flight. It's a sight that inevitably magnifies their bravery, skill and verve. It's hard to imagine you could ever tire of this ringside exhibition of sheer horsepower, let alone cease to be amazed by the touch and ballet-like agility of the riders as they shift, shimmy and lean on their bikes. [1] Chris Moss drinks in the skill of the riders, "The g-force which they counteract to keep the centre of gravity is just amazing!" Ian Smalley's a patient guide to the intricacies, "What you notice from here is the tiny little changes in position that can make such a big difference."

The next race sees all four riders skitter off their bikes in the first corner. To untutored eyes, this looks a real conflagration. Ian quickly reassures, "Basically that's called first bend bunching!" As if to emphasise the difference between a proper crash and a tangle, heat 6 sees Lee Richardson and Maciej Janowski lock together on the apex of the first bend and proceed to hit the airfence with great velocity. Though the safety furniture does its work, Maciej disappears at great speed under the skirt of the airfence. Unlike Lee Richardson (and also Cameron Woodward), Janowski then spends a good time prostrate on the stock-car surface surrounded by paramedics. Eventually he gets gingerly back onto his feet. Chris and Rob rightly marvel at his hardiness. Ian explains, "That was a proper crunch! In the Elite League they have air fences but, in other leagues, it's often wooden. The riders have lightweight protective body armour and also wear race suits made out of very strong material that are called kevlars."

If the Time Out team's flabber was gasted by the noise and impact of the crash, actually seeing Maciej Janowski return to the start-line for the rerun is also notable. That he then wins verges on the remarkable! Photographer Rob savours the centre-green thrills, "effing hell, that's just brilliant!" Obviously, all experienced speedway types affect total nonchalance at Janowski's return and subsequent win. [2]

To the first-time visitor, race results are an irrelevance. Drawn heats, track records, tactical rides and disqualifications pale into insignificance compared to the elemental speed, dynamism and harum-scarum derring-do of each and every lap. How the riders negotiate each bend absorbs and enthrals (for me, anyway!). Ian patiently explains, "The corners here are wider than people think and the banking is much more than people know!" With his marketing manager's hat firmly in place, Ian underplays his own role in the Royal Navy club sponsorship preferring, instead, to describe the structure, dynamics and importance of the speedway sponsorship proposition.[3] "Basically with an airfence banner on Sky, we can sell these at a premium! Obviously, certain spots on the track are more dangerous than others – they're all dangerous, of course – but we have all the stats and the prices reflect that!" Ian succinctly discusses the club's communication strategy ("We try to engage with the fans and be open and honest every week"); the Speedway Grand Prix Series ("The GPs are the great product from a broadcast point of view but they do them in blocks of four so the stops and starts work against them") and the riders ("It's a self-employed profession"). Lakeside's linkup with the Royal Navy interests the visitors. "I got the Navy involved through a contact of mine. What the Navy get for that three-year contract is mass marketing advertising on all our platforms, which gives them more awareness for their point-to-point demographic than a £25,000 ten-second advert on Sky would! They get their branding on the air fences and pit boards and also they had more direct hits here than they did at Silverstone in three days!" Chris and Rob react like youngsters in a sweetshop and automatically gravitate towards the start/finish line area.

[1] It clearly impresses Chris Moss if judged by his article. "I go to stand at the starting line. Here the fix is different: it's all about noise. The screaming hellishness of it is wonderful, and I watch in boyish awe as the four riders shoot off, pulling wheelies for 20 yards before they lean right down into the bend and accelerate out of it – wheelies again, jostling for position, dodging and ducking to confuse the competition. Sparks fly off metal boots as feet strike the track to keep their balance.

[2] In another section of his article, Chris writes, "In one race all four riders hit the deck on the very first bend. In another, a lone biker smashed into the so-called "airfence" of protective cushioning material that surrounds the track. The thud, however, sounds like a wooden mallet hitting a skull. I assume he's dead. He gets up, gets back on his bike and they start the race again."

[3] Again from the Time Out article: "He's making progress", [Ian Smalley] says, helped a lot by the fact that Sky Sports now broadcast speedway on Monday evenings and thanks to an ongoing sponsorship deal with the Royal Navy. "I think of speedway as the extreme team sport for all the family" he says "and, look, these are ordinary people here in the pits, meeting their heroes. You wouldn't get that in football or Formula 1: most sports have become so remote."

The start marshal with the distinctive chef's trousers – Ian Glover – chats amiably (after he remarks out the starting gate positions). Even with the new silencers, the roar and vibration thrown off by a speedway bike when revved to extreme seizes your insides in elemental fashion. Between races Ian Smalley makes no attempt to hide his inner speedway fan. "When Poole are here – and you only get it with Poole and you get Darcy and Chris – you can hear them whooping after they finish! And when they won heat 13, you could hear them screaming down the back straight. They are something special! Speedway riders are a unique breed!"

The spectacle's so wonderful out on the centre green, Chris Moss wonders, "Why can't they have people sit out here on VIP seats? It's not dangerous here."
"Yes it is!"
"Why?"
"Well when a bike flies over at 40mph, you're not gonna be able to dodge it!"

Though ignoring the white tyre rule, both Rob and Chris love to stand close to the tapes as the race starts. Fortunately, tonight's SCB referee Mick Posselwhite takes a relaxed attitude to this technical infringement of regulations. Chris drinks in the power, vibration and excitement. Both journalists think the noise is incredible. Camera in hand, already Rob has a greater appreciation of the action, "They rely on all the other riders. It's the straights not the bends – it's the acceleration down the straights." With a Press Officer to hand, Chris asks more technical questions, "How much do the bikes cost?" "Without tuning and other bespoke tweaks and additions", Ian replies, "£3,500 before they do anything to them!"

Rob and Chris enjoy it so much that they watch eight heats (4 to 11) from the centre green. Between races, Chris speaks to all and sundry including Community Manager Mark Sexton ("It's hard but rewarding work. You do have to give it away to get something back sometimes.") Start marshal Ian Glover passes on some key advice, "It's best not to tell anyone how much better it looks from here!"

The Time Out trip to the speedway wouldn't be complete without a burger. While Chris queues, Rob gets out his mobile "I'm calling my dad."
"He's missing out."
"He lives in Birmingham."
"He can see some speedway there."

After a meeting won comfortably by Lakeside, 43-year-old Rob remains enervated by the thrill and excitement of his speedway experience, "I can't tell you how impressed I am! I just feel really angry that I didn't go to Wimbledon now!"
[Ian Smalley] "It's nice to hear that because – Jeff knows – you bring a lot of people – but there's so much choice these days."
[Chris] "Are you really happy with that result because it's kept your Play-offs hopes alive?"
"Yeh!"

**12th August Lakeside v. Swindon (Elite League) 56-38**

CHAPTER 22

# Swindon

## "Tyres take ten minutes to fit and silencers two minutes!"

8th September

Ronnie Russell

After a disappointing season by their standards, Swindon's last home Elite League fixture of the season sees Peterborough Panthers race at the Abbey Stadium. Though nigh on invincible at home, the Panthers have yet to win away this season. In any season to qualify for the Play-offs, you need away wins. Despite Swindon's stuttering home form, they start favourites – especially since Tom Armstrong rides at reserve in place of the Panthers Kacper Gomolski. It's a contest some see as the speedway equivalent of two bald men fighting over a comb.

At the Abbey Stadium entry turnstiles, long-time Robins stalwart Bob Fisher is affable and well versed in the ways of the speedway world. "It's been a poor season here. It started off badly."
"With Leigh?"
"Well, he was always going to be impossible to replace and leave a big hole in the side. Then there was Zengi getting injured. Oh, and the uncertainty with Coventry and Peterborough. Well and also the Poles not knowing what they were doing with the silencers – that delayed the start of our season for a couple of weeks. So it was stop and then start and we never got going! I was saying on the way here, if you look at the team on paper we're excellent but we've been inconsistent. Inconsistency has been our problem. The racing has been very good but not the results. If you talk to some fans, they don't like it because they want to win all the time, but speedway isn't about that! I've been coming 50 years – and I suppose I'm getting cynical – but it's the excitement we come for really! Last week Nicolai Klindt was brilliant against Harris and Larsen. They were on a 5-1 and Harris started to get nervous and hung back and Nicolai passed him and then sailed straight past Larsen. I was shouting and screaming – like the rest of the stand – it was brilliant but then – in heat 15 – Nicolai was last. He started with the rest of them but finished nowhere. Steady was out for a while and when he came back he'd missed so much he couldn't catch up. The silencers business was silly. They should do what they did for the tyres – supply them at the track. Tyres take ten minutes to fit and silencers two minutes! You want good racing every week – like you get in the

GPs or in the British final. Look how Scotty rode against Bomber – really hard but Bomber didn't mind because that's speedway. He'd have done the same! I don't know if it's because I'm getting old but I enjoyed Sky to start with but now, they shout and scream all the time. They do that in all the sports not just speedway. They want a Poole/Coventry final again so Nigel and what's his name can scream and get excited."
"Can't Peterborough still get there?"
"Not after last night. They can't catch Lakeside and King's Lynn is now above them so it's all over."
"Maybe people will come to see this new Dane [Mikkel Bech Jensen] they've got tonight?"
"Did you hear Emil's manager has said that Emil won't be riding after Monday? He was never going to! If Coventry gets to the Play-offs, they'll have guests! Which isn't representative of their team. It isn't right, is it?"

Swindon club historian and prolific author Robert Bamford wanders over with Dave Stallworthy. Dave used to announce at Reading speedway (and wrote a book on the club) but nowadays is in great demand as a toastmaster at various social events where Master of Ceremony skills are required. Bob Fisher modestly takes his leave of this gaggle of authors, "I won't interrupt the literary circle though, a few years ago, I did a column for the programme every week". Robert Bamford wonders, "Do you think Scott [Nicholls] will be heckled at the end of season when they have the rider speeches?" Bob isn't so sure. Rob isn't either, "He's been awful Bob, he's been bloody awful. The only rider worth his salt this season has been Nicolai Klindt!" This seasons performances and speedway politics generally have, after many decades, worn down Robert's enthusiasm, "I'm coming to the Bob Kilby meeting and them I'm going to retire from the sport!" Talk soon turns to the sadly defunct Reading Racers. Dave Stallworthy needs no second invitation to wander down memory lane. "I was an announcer from 1996 to 2006. For 11 years. I did a quarter of the time Reading was around!" Ever the historian, Rob interjects, "Reading was open from 1968-2008." "I was doing the National Vintage speedway championships on the Sunday – they had them for four years I think it was '88, '89, '90 and '91 – and Pat Bliss heard me and said 'come along tomorrow!' It was the end of the 1988 season and I think it was Jan Andersson's select – I think they called them Jan's Mixtures versus Malcolm's something or others (at Jan Andersson's testimonial meeting)-– and Bob [Radford] said he was off to the bar and I did it and I was relief from then on in!"

Since Reading Racers are my team, I can genuinely say to Dave (that along with Dave Lanning, David Hamilton, Paul Hunsdon and Bob Radford) "You were one of the voices of Reading Speedway for me!" "For me it was a great honour. I followed other greats like Dave Lanning, David 'Diddy' Hamilton and Bob Radford. Oh, and I enjoyed working with Peter York. When I started doing it in '96, the atmosphere was poor so I tried to build it up again. There were some girls who used to do some cheering – you probably heard them? My only involvement nowadays is doing the commentary for the Plymouth Devils from 200 miles away! I still went to Swindon so I had a blue-and-white bobble hat for Reading and a red-and-white bobble hat for Swindon. I've done over 80 weddings this year on Fridays, Saturdays and Sundays so, even, my grasstrack has had to stop."

Over at the freshly laid trackshop tables, there's a new addition to the team. Ellie lies in her pram closely supervised by proud new mum Lucy Aubrey (father Lee Poole is elsewhere). "She's nearly four months. I'll be here for about an hour. She came to her first meeting last week but left just before the meeting started. She wasn't bothered by the bikes and, when we're at home, she likes the speedway and the Moto GP."

Inside the Speedway Office, affable Wayne Russell is hard at work and hopes the latest Swindon promotional campaign is hitch free. Wayne stuffs envelopes with £10 Swindon Speedway entry vouchers for the Bob Kilby Memorial Meeting on Thursday September 22nd. "Every fan here tonight

will get four which hopefully will bring lots of people along to the Bob Kilby memorial meeting." [1] Unlike many managers in the speedway fraternity, Wayne Russell not only acknowledges that he looks at the British Speedway Forum but, surprisingly, is happy to reveal what he enjoys, "I like Steve Shovlar – it's all tongue in cheek to get a reaction – isn't it?" Wayne is philosophical view about a difficult season for Swindon and prefers to praise the achievements of others. "It's good to see Eastbourne in the Play-offs and King's Lynn."
"It would be nice if Eastbourne could have two meetings with four or five thousand people."
"It'll be more like three, I think, but good for them! It will be interesting to see who Poole chose? I reckon King's Lynn because Chris, Darcy and Davy will all fancy their chances there – it could be very interesting!"

In the pits SCB official, Margaret Vardy performs her various duties while partner Steve Davies (resplendent in a white replica Walsall top) soaks up the atmosphere of an empty stadium. Out on the track, the curatorial team – made internationally famous by Graham Cooke's Blunsdon Blog – industriously go about their last-minute preparations. Steve wonders, "Wouldn't it be cheaper for you to pay to get in rather than writing a book to do so?" Though bathed in sunshine, Steve has meteorological concerns, "What about those dark clouds?"
"They'll blow over."
"I hope so, I've had four hours' holiday to see this."

Away from speedway, Steve is closely involved with Walsall. "I've become Walsall's Supporters' Trust secretary for my sins. We had these Supporters versus the Legends football match last Sunday and the Legends team included David Kelly. Our charity this year is SANDS – which stands I think for Stillbirth and Neonatal Death Society – because Walsall club secretary Dan Mold and his wife Emma are aiming to raise £10,000, so it's also our cause for this year. [2] We've got band nights and all sorts sorted for the rest of the season. We had our first match last week against the Dudley Heathens. We won 3-2. We were always gonna win because I was the ref. I believe Mark Robinson played. He's the 43-year old ex-Cradley rider – he was only reserve in days gone by. He's going to make his return debut in the sport for the Dudley Heathens shortly. I wouldn't know which one he was. Dudley Heathens team manager Will Pottinger scored an absolute screamer – 20 yards into the top corner. At the end of the speedway season, we're going across to play their riders because they can't play in the season. It would be good if we could nurture something with the Heathens. The thing is Walsall don't have a speedway team and our fans would never support the Wolves – they're mostly West Brom fans, anyway."

There's increased talk (in the press) that a new stadium is to be built in Swindon to replace Abbey Stadium. Steve has concerns, "It would be a shame if this track goes. It's a shame when any track goes but this is a bit of a one-off these days. There's not so many big fast racetracks anymore! Okay, there's the Isle of Wight but they've lost Exeter. Another problem is though, there's lots of characters in speedway, they're all old or getting old. I don't suppose you've been in the grandstand when Birmingham get a 5-1? It's pensioner aerobics! They put the music on and they all get up and go for it!"

Wearing impressively worn overalls, Graham Cooke interrupts his curatorial promenade on the tarpaulin dog track to discuss his post-meeting plans. "GB Team Holta are going to say goodbye to Rune. We're going to Vojens on Saturday and the Danish Under-21 final at Grinsted on Sunday and then fly back. According to Jan St  chmann it's a very black and dusty track if you stand in the

[1] The voucher terms and conditions read, "Please use one voucher yourself and pass on any others to friends and family that you feel may like to come and watch speedway for just £10. Remembering a Swindon Legend!" In order for each £10 entry voucher to be valid: fans must complete their name, email and mobile phone number.

[2] The charity supports anyone affected by the death of a baby. They work in partnership with health professionals to improve the quality of care and services offered to bereaved families and to promote research and changes in practice that could help reduce the loss of babies' lives.

wrong direction. Is this your first time to Swindon this season? You haven't missed much, though we've had some excellent racing! Nicolai Klindt has been the star. The guy just doesn't know when to shut off. Last week, he had a brilliant race with Harris and Larsen. You could see the airfence rippling as he went around! It's sad but it's all gonna kick off again in the close season. Crouch [Dave Croucher] isn't here so number 96 looks like a bombsite." BBC Wiltshire speedway journalist and club announcer Richard Crowley interjects, "I don't know if I've enjoyed myself this season? Well, there was that race last week."

[Steve Davies] "There's a positive – that brown bit out there!"

[Richard] "That's true."

[Steve] "Look how delighted the Leicester fans are. I'm sure if I'd posted something about how bad things had been at Swindon – fans from Crewe, Cradley, Ellesmere Port, Reading and Bradford, who'd all welcome a track – would get in touch saying 'get a grip'. I'm sure there are lots of fans from other defunct clubs who'd envy what you have here!"

Motorcycle News sports editor Gary Pinchin is at the Abbey Stadium as part of a whirlwind mini speedway track tour. "I've found your Track Directory very useful on my tour. This week I've been to Poole, the Isle of Wight, which was rained-off, where did I go last night ... ah, yes, Somerset and now here. I used to live just down the road in Devizes but now I live in Lincolnshire. This was the first track I ever came to! I have a blog and I wrote on there I was following your example on my speedway tour. Shall I tell you what I thought? Poole was really good! The Isle of Wight was full of puddles and Somerset – it was the first time I'd been there – it was absolutely fantastic! Wonderful even though it cut up badly, there was still some brilliant racing. I live in Stamford so Peterborough is close. In fact, it's outside our office but it's not very exciting. I went to Scunthorpe, that was dreadful but I kept thinking there must still be some good speedway somewhere and I found it at Poole and Somerset. I like the back-straight grass bank there – this isn't work, it's pleasure! Tell you what got me started off again, how this all started. I went to Sparkford Haines Museum because I had a job down there and I didn't realise there was a Speedway Museum. It's the Forshaw Collection. [3] There's a link to it on my blog. I've been to a couple of Peterborough meetings but haven't really enjoyed them. A guy said to me last night at Somerset, 'Who do you support?' and I said 'I don't, I like to go and see good racing!' Even when I came here a lot I didn't support Swindon. I used to go to Exeter as well. I'd catch the train down there but there was never a train back, so I used to sleep at Bristol station. I used to go to Newport too – the old Newport, Somerton Park. I reckon it was the best track ever, Crumpy was magnificent round there! Newport was a weird racetrack with a double apex corner. Then Bob Kilby and Martin Ashby used to go to the outside here and cut back. When I got your photo book [Shale Britannia] I tossed it to one side thinking 'what the fuck is that?' Then I picked it up again and I thought it was brilliant!"

Over recent years, any visit to Blunsdon isn't complete without the presence of ex-Antiques Roadshow worker Harold Davis (91) and his friend Anthony Roberts (aka Rob) stood by my book display table. Tonight Rob's here alone. "Harold has had a stroke and he lost the use of his leg and hand. He's in Frenchay, it happened a couple of weeks ago. He's gradually using his leg and his hand but, when you're 91, it takes longer, doesn't it? Would you like to sign his Get Well card?"

The Sittingbourne massive – aka Dick Jarvis, Chris Golding and Les Drury – are on their own mini speedway tour that saw them at the Oak Tree Arena yesterday and Blunsdon tonight. Hans Andersen's arrival in the Peterborough Panthers team for Nicki Pedersen is a significant lure for some fans but isn't for the Sittingbourne massive according to Dick Jarvis, "I thought we'd better come in case they don't run next year." Robins Supporters' Club organiser Darcia Gingell gives a quick review of the season to date. "We've been talking about next season since about June. A few

[3] In the Forshaw Collection they have "31 rare and interesting speedway bikes".

years ago if we lost once at home it was a disaster but, now, we shrug it off all the time. There's been some good racing but the main thing is that we have a track to go to!" Further conversation is cut short by a sudden fanfare of loud music blasting out over the stadium speakers. Roughly calibrated at kill level, when the chorus of 'When the red red robin comes bob bob bobbin' along' fades, Clive Fisher – apparently under the mistaken impression that he's the MC for a Rolling Stones concert – shouts intensely into his centre-green microphone, "Good Evening Swindon!"

This joyous scream is, in fact, the cue for a rather bouffant Robert Bamford to retrieve his box from underneath the trackshop tables. He'll mostly stand on it for the duration of the meeting. After their ceremonial parade truck circuit of the Blunsdon track, both sets of riders dismount for some pro-forma introductions from Clive Fisher. I'm surprised to learn Linus Sundstrom is Panthers captain and also have a rider called Tom something-or-other at number 7. Clive obviously knows Tom's surname but the sudden onset of a regional accent as well as his failure to enunciate clearly makes figuring out who actually rides difficult. It sounds like it's "Jenkinson". Dick Jarvis tells me it is, "Tom Watson, he's an Aussie – he rode a few weeks ago at one of those Birmingham Bulls matches."

Peterborough's number 2 Michael J. Jensen wins heat 1 comfortably enough though the real battle is behind him. Scott Nicholls welcomes Hans Andersen back with a big shove on the first bend. No sooner does the race finish than referee Margaret Vardy sounds the two-minute warning. Richard Crowley immediately says, "That's a quick two-minute warning perhaps because of weather moving in later." Dick Jarvis reveals they also suffer from weather at Iwade. "The other Sunday it lashed down but we left it half an hour and didn't touch it and it was alright. We coned off the inside line but, after an hour of racing, we were watering it." Dick isn't surprised some promoters are slightly oversensitive about my written work. "I don't know why promoters get funny! It's only one person's view. What would they do if a hundred fans gave them their views of the previous season? Would they ban them or threaten them too?"

Carrying on tonight's tradition of foreign riders called Michael Jensen winning, Mikkel B. Jensen does the honours and celebrates with a slightly superfluous – given the absence of North and a fall for Armstrong – but exuberant wheelie over the finish line. Easily impressed by any Robins rider race win nowadays, Clive Fisher burbles, "A tremendous ride from Mikkel B. Jensen in heat number 2!" Robert Bamford rebuts my suggestion that referee Margaret Vardy should disqualify both Swindon riders, "So you think that wasn't a bona fide attempt to race then?" Dick Jarvis marvels at the size of the crowd. "I think this is quite reasonable for a team that's bottom of the league! Still, it is the last league meeting of the season and, I suppose, the other side of the coin is it could be the last league meeting ever so they thought 'I'll come and have a look.'" Some disgruntled Robins fans claim that Jesper B. Monberg has subjected them to a season-long campaign to reduce his average. They won't have their minds changed in heat 3 after his initial brio quickly dissipates into a last.

Heat 4 provides a battle of thrusting young Australians – Cory Gathercole and Troy Batchelor – with Nicolai Klindt thrown in for good measure. Tom Armstrong soon retires and the Aussie duel ends with a dramatic tumble into the bend-four safety fence for Batchelor. Uninjured but soon disqualified, the race is awarded as a 5-0 to Swindon. Those in the know at Swindon are (possibly prematurely since he only gains one further point from his next three rides) delighted with new signing at number 6 Mikkel B. Jensen. Rob Bamford pronounces, "He seems to be a good signing" while (post-interview) Clive Fisher remarks, "Isn't his English good! Better than my Danish anyway!" Richard Crowley corrects, "Better than your English, Clive!"

During the last lap of heat 5, after biding his time, Peterborough's Michael J. Jensen overtakes teammate Hans Andersen to snatch third behind Messrs Stead and Monberg. Moans about average manipulation by Monberg temporarily evaporate. With kevlars reputedly five times tougher than steel, Troy Batchelor demonstrates his own resilience after his big heat 4 fall, by winning heat 6. Whenever Dakota North's on the track, things are never dull. He gets out of shape in virtuoso fashion but somehow stays on his bike to finish fourth. Already 11 points to the bad, Panthers team manager Trevor Swales sends out Panthers captain Linus Sundstrom on a tactical ride in heat 7. Peterborough's Krzysztof Buczkowski (aka Buzz) leads from the off and, until almost the finish, appears oblivious to the tactical need to let his teammate Linus past. In reality, what appears like obliviousness is, in fact, race smart and shale savvy since slowing too early would allow the fast-charging always-there-or-thereabouts Nicolai Klindt through for a (disastrous) race win. Prior to heat 8, Armstrong and Jensen line-up in the wrong start positions before the young Dane zooms off to win comfortably. From a Panthers perspective, heat 9 is something of a farce since Troy Batchelor is disqualified under the two-minute warning and then fails to make it to the start line for the rerun. Stead and Monberg then proceed to make heavy weather against the solitary Peterborough rider Dakota North. Dick Jarvis doesn't hold Trevor Swales' tactical nous in high esteem, "Impressive team management from Peterborough not putting out North in heat 8 and then not replacing Batchelor in heat 9 or putting him off 15 metres!" After Scott Nicholls crosses the line to win heat 10, "Beautiful Day" blasts out at ear-splitting volume over the stadium loudspeaker system. Richard Crowley interrupts these musical celebrations, "Message for the riders who've left to do the Speedway Riders' Benevolent Fund collection – could you please return to the pits now!"

If there is a collection during the interval, I don't see it. Heat 13 is the most exciting race of the meeting for the duel between Nicolai Klindt and Hans Andersen. It's the only race where Hans looks anything like the Hans Andersen of old (with an outside swoop past Klindt). Simon Stead's run of three races, three wins ends with a fall in heat 14 while, up front, the 5-1 from Sundstrom and North raises the possibility, if repeated, of a drawn meeting. Dick Jarvis shrugs, "Swindon's going to snatch defeat from the jaws of victory!" It's a forecast of doom that doesn't happen since Andersen is last with Sundstrom third. On his way out of the stadium, Gary Pinchin exalts at the quality of speedway at the Abbey Stadium, "That was another great meeting! When I said Peterborough were rubbish, I didn't mean the riders! Buczkowski was brilliant!"

After the victory parade, Clive Fisher conducts post-mortem analysis of the 2011 Elite League season with the various Swindon riders. Cory Gathercole has a soothing voice but is a reluctant interviewee, "Thanks for sticking with us all season – we couldn't pull it off!" Simon Stead's much more practised. "It was good to finish off with a few heat wins. Thank you all for your tremendous support this season, it means a lot to us! Speaking for the team, I'd like to thank every one of you from us all! We can build on tonight – there's a lot of positives. Mikkel Jensen's a good Danish prospect and we're building for the future and I'd like to be part of that!" There's a finality to Clive Fisher's opening question to Scott Nicholls, "Well Scottie, it's been great to work with you." Scott admits it's been "frustrating" before he continues, "we've had a lot of bad luck ... I've had a dreadful season – not for the lack of trying! It's fantastic that a lot of people keep coming back each week. It's no consolation but, at least, we sneaked a win and got a small amount of pride back. After most of the meetings at home, I wanted to hang me head but I'm building like mad for next season. I've no idea where I'll be next season – it's my job – but I do the best I can to remind myself why I do this job and enjoy it. I've had my criticism which I take on the chin because – without you people – we wouldn't have a job." Clive wonders if the close season will again be characterised by protracted turmoil. Scott replies, "Hopefully they can get their heads together to plan for the future. If I'm not back here anymore, thank you for having me here!"

With some post-meeting practice laps still in progress, the crowd stream away. Season ticket holder Jim tells me, "I'm buying a book to celebrate the worst season we've had and I'm a season ticket holder!" Robins quiet but optimistic incident recorder (and ex-Reading timekeeper and incident recorder) Roger Nettlefold asks, "Do you have any news on Reading?" Sadly I don't but I'd love them to return. I'm sure Swindon fans would like more detail about the promotional team for next season as well as concrete news of future stadium plans. The security man looks onto an almost empty stadium. "If I wasn't being paid, I don't think I'd have come to this season. I remember when we had to ask people to leave the stadium! Then, after, we'd all go upstairs and have a drink and curry – about 35 of us – riders, staff and fans. Those were the days and they're not coming back!"

**8th September Swindon v. Peterborough (Elite League) 51-43**

# CHAPTER 23

# Sittingbourne

## "I set off and something went whoosh and I thought, what's that?"

12th September

Trophies

I'm perhaps a fraction biased but – apart from Buxton – where else in British speedway could be better for a round of the 2011 British Youth Championship than the Old Gun Site in Raspberry Hill Lane in Iwade, near Sittingbourne? The fifth round of the 500cc, 250cc and 125cc championships takes place in front of a small but dedicated keen crowd of fans, helpers, friends, family and fellow riders (whether junior, current or retired). Also on hand is the additional expert advice of Phil Morris and Neil Vatcher as well as SCB referee Jim McGregor. With 30 minutes to go prior to the noon scheduled start time, Jim calls a pre-meeting briefing for riders, mechanics and helpers. "I don't want you digging at the start for longer than you're racing! If you all sit still at the gate, then I won't hold you but, if you are moving, I'll hold it! Remember your bikes aren't as highly tuned as the professional riders. If I pull it back, just come straight back round to the start line. Wear proper helmet colours. We've got a lot of races to do – 32 races and the grading – so please keep a move on! Any questions, lads?"

Amongst the volunteer staff at Sittingbourne speedway, there are many familiar faces including training instructor Paul Heller who, unsurprisingly, reveals the club cuts their cloth according to the financial environment. "A couple of the training sessions we've run haven't been so clever. Three on the mini track and six on the main one means we've lost money. We still have to pay all the fixed costs so we've scaled it back a bit! During the season, everyone can get lots of rides all over with second halves and that. But – during January, February and March – we can have 40 or 50 people queuing up here to ride because everyone is wanting to practise, get ready for the season or test things out!" Well-travelled Bryn Williams is our meeting announcer. Usually phlegmatic, sometimes he registers frustration. "There's lots of idiots on the British Speedway Forum. There's 63 pages or something like that on Coventry or is it 100 pages? As soon as Allen Trump leaves the sport, things will calm down. Tony Lethbridge at Exeter doesn't usually swear but he does about Trump. They do at Oxford too!"

Keen as mustard to perform well in the 250cc British Championship, Eastbourne fan Jordan Hazelden has been hampered by injury but this afternoon returns to the track. All his family – parents, grandparents, aunt – along with some friends are here to watch. His father Mark remains studiously philosophical, "He's very nervous. I've told him to go out there and enjoy himself but it's easy for me to say. He hasn't ridden, it's his first meeting back from injury and he's got an incredibly tough heat 1 against Kelsey Dugard, Reece Downes and Nathan Greaves who's leading the championship!" Passionate lifelong Eastbourne fan Judy – Jordan's grandmother – waits anxiously for the action to get going. "I shall be glad when the meeting's over to be honest. It's his first meeting back and it's the first time I've seen him ride in a competition. He's pretty sensible but, unfortunately, you can't predict! Put it this way, everything is crossed!"

Prior to the rider parade, Rerun DVD's supremo Steve Girdwood rests with his 4-year-old son Tai on his lap in the clubhouse. Since he was a babe in arms, Tai has been to the speedway, "Dad, when the traffic lights go red the riders have to stop." "He knows a lot this one! He said to me the other day, 'I can't see the sense in that, Dad!'" Steve's work takes him to a wide variety of British speedway tracks every week. Last night Eastbourne qualified for the Elite League Play-offs for the first time since 2005. Ignoring his vested commercial interest, Steve is delighted, "I kept thinking the bubble must burst with Eastbourne but they keep going and could win something!"

While we wait for the arrival of the paramedic, 29 young riders troop out onto the centre green for the rider parade. In a lilting Welsh accent Phil Morris introduces each rider individually with words clearly based on affection for and close knowledge of these young riders. His comments include Kelsey Dugard ("He's got a very good pit crew – Martin Dugard and Martin Hagon"), Josh Bailey ("He's a little lad – don't let that fool you – he puts it about on the track") and Macauley Leek ("This lad seems to be stuck to my leg or Neil's leg asking if he's in the final or the next race"). Understandably, some riders are on the 'diddy' side except for the tall lad ("He's only 10 too!")

Erstwhile Mildenhall promoter Steve Ribbons is here and greets me with the trademark honesty so beloved at the BSPA, "You're looking fat! You've put on a bit of weight! You get the truth from me." Though Steve claims to be philosophical, some aspects of the handling of the rejection of his application to run Mildenhall speedway continue to rankle. "I sent a formal letter and haven't had a reply or as much as athank you! I had a fully laid out business plan. I'd not be going round with a hat like lots of speedway clubs and – though, apparently, they had money worries – Yet, I wasn't even asked about my finances. I have a quarter million pound haulage business, me mum died and left me a property and I've a hundred thousand in the bank so there would be no finance worries. In fact, the business plan they're using at Mildenhall now is the one I'd written out with Training Schools and that, that they now call an Academy. We weren't just going to hope that people turned up, we were going to actively promote it! For example, we were going to have a 40-foot banner on bends three and four saying something like - THIS IS MILDENHALL SPEEDWAY! So – when you get in the stadium – you get the wow factor as soon as you walk in. Peter Morrish told me I had to pay £5,000 to be a League member, £5,000 for promotional rights and that's in addition to the £3,500 for the Bond, which I knew I had to pay. And, although lots of other promoters don't have to pay these, I reasoned it was worth a go, after all I only got involved because I didn't want to see the place shut." Steve pauses for effect and then grins, "And that's despite being a Rye House fan! After that I said I would pay I was told - 'you can't come to the meeting but Kevin Jolly can have an audience'. I don't understand it? I had a promoter's licence in the past, which wasn't revoked or rescinded? Yet I still have never been told what I've supposed to have done or been accused of! I'm not sure if Peter Morrish did that off his own bat or made it up as he went along. Though I'm not happy about it, I can't get away from my inherent enthusiasm for speedway. From 1978, I've stood out on the terraces. A lot of them who were there then have drifted away! I dunno why, but I can't give up on speedway. I'm always trying to think of new ideas. We all know that the Danish riders come over here on tour but has anyone ever been over there on tour? I've been chatting to Kelvin Lapworth, the Danish League has a break in August and they have third division clubs that are crying out for people to go there then. You know, there wouldn't have been any money worries if I was at Mildenhall, which I believe, is now guaranteed by the Ipswich promotion. You know, I think I would have rather enjoyed it!"

When the paramedic arrives, the meeting gets underway 40 minutes late under the tutelage of the Young Lions management team 2011, Phil Morris and Neil Vatcher. The riders set to the task in hand with enthusiasm and some panache. Second cumulatively in the 500cc championship, Nathan Stoneman wins comfortably and won't face the championship leader until the 28th race this afternoon. Heat 2 is the first 250cc race of the afternoon and features comeback rider Jordan Hazelden. It's awarded after Nathan Greaves falls on the second lap. Out in the first 125cc race of the afternoon (heat 3), Somerset's Henry Atkins looks composed and stylish on his bike. He's an affable young man who's a mascot at the Oak Tree Arena. Michael Neale's afternoon (500cc championship) starts uncomfortably after he loops immediately at the start and writhes in agony on the shale about 10 yards further forward. A man I suspect to be his father rushes over from the pit gate screaming loudly, "Whoa, whoa, calm down." Michael initially fails to heed these words and we also hear the lady paramedic remark, "I thought he was telling me to calm down!" Disqualified from the rerun, Michael's escorted back to the pits by his dad and Phil Morris ("getting a little bit of coaching from the whizzer").

The second 250cc race of the afternoon (heat 5) goes ahead without the evocatively named Zach Wajtknecht (Bryn tells us "he's broken down en route"). The initial attempt to run this race results in a disqualification for Josh Bailey and a fall for Jordan Hazelden which, judged by the pained expression on grandmother Judy's face, is exactly what she didn't need to see. Rather impressively, Jordan's dad Mark actually runs the length of the centre green to attend to his fallen son. After some TLC for rider and equipment alike, referee Jim McGregor sounds the two-minute warning, ("primarily for the rider in red"). Sadly, though Jordan makes it back to the start line, his engine failure leaves Connor Mountain to enjoy an untroubled solo four-lap race.

While we watch the track grade, Steve Girdwood recalls his speedway racing career. "I did the mechanic races in 1989 and 1990 at Ipswich. I didn't take it that serious. I had flip-flops stuck either side of my helmet and on my back was 'See yah' that none of the others saw! I was known as "Gater" Girdwood. I'm still known as that at Ipswich! In my first race, I gated for about a yard and that's how I got my nickname! I can also say I've raced against a speedway world champion, Mark Loram (though he hadn't won it then). I raced him in a small little track in South Africa called Corobrik in Johannesburg. It was three laps for me and five laps for Mark. There were 200 people watching so I really got quite nervous. We were racing for a Castle Cup lemonade bottle made into a trophy. It was such a small track, the racing was like Costa Mesa. Unbeknownst to me it was all filmed and Neville Tatum did the commentary. I set off and something went whoosh and I thought, what's that? It was Mark. I didn't know it but he'd hit a hole on the third bend – they'd done nothing to the track from the racing the night before – and he came down and dislocated his thumb and had to go to hospital. This was 1989 and the *Speedway Mail* reported it. Because of the injury, he might miss the start of the season but they said something like 'he was obliging some fans by posing on a bike when he hurt it' rather than say he was racing me! So that's it really, I beat a future world champion! So I do class myself as a retired international speedway rider but never did write that on my van. In 1991 I then rode against Noddy Naude's wife off scratch and I lost, so I retired. It was the time of my life going to South Africa. I went for ten weeks in 1989 with Chris Louis, Mark Loram, Kelly Moran, Richard Knight, Andy Galvin, Neville Tatum and Rob Tilbury. It was fantastic. I filmed one meeting a week in Johannesburg and we went to Durban four times. What I made from the video sales paid for my holiday, except the flight. I was 28 in 1989 and had only just got into speedway. They were great times and the stories I could tell! It was all organised by Pete Murray, the ex-Wimbledon and Canterbury rider. I went there for six weeks in 1991 as well. Looking back, I suppose where we were staying was a bit rough but we didn't know that at the time. When we arrived, we went straight to the track and when people asked us where we were staying we got some funny looks that we didn't understand but we did shortly afterwards. I won't tell you the stories but they were brilliant times. One night there was a gun went off in the pub we went in – I wasn't there. I think it was the only night I didn't go to the pub on the whole trip!"

The third 500cc race of the afternoon (heat 7) only gets as far as the third bend. Jamie Couzins falls and, with lightning reactions, both Danno Verge and Jack Jones expertly lay down their bikes to avoid undue carnage. Though these riders are taking their first tyre turns within speedway, there's no lack of ferocity out on the track. In the rerun, Nathan Stoneman certainly doesn't take any prisoners and passes Danno Verge with great aggression. While the third 125cc race of the afternoon (heat 9) roars round the track, Graham Arnold hovers by the shoulders of the Baseby brothers, "You old boys were out on these, weren't you?" SCB track inspector Colin Meredith is at the meeting albeit not in an official capacity. His revolutionary style safety fence is now in use at the Norfolk Arena. "I just want to make speedway safer and, if that upsets Tony Briggs, tough! A monopoly is no good for anyone. Birmingham, Leicester and Wolves are all interested. We need them everywhere, it's got to come."

The fourth 250cc race of the afternoon (heat 11) piles on the agony for the extended Hazelden family after Jordan falls heavily on his second lap. His bike also hits him (on the back). Though Jordan stands and clears the track, he collapses on the centre green. His father Mark is again quickly out there, Martin Dugard trots over at a pace that – in cricket – you'd call military medium and his granddad John strolls over. Grandmother Judy tries to look on the bright side, "Winded him, I hope that's all it is!" With complications to his recent injury uppermost in minds, Jordan walks back towards the pits to a polite ripple of applause. Martin Dugard and Neil Vatcher push-cum-pull his wrecked bike. John soon updates Judy, "He's hurt his back and got a red swelling on his head and the bike is a write-off but, at least, it's not his leg like I thought it was!"

Inside the canteen, a lady breastfeeds her baby, "I think the speedway knocks her out – I wish I had speedway at 2 o'clock in the morning!" Contrary to his reputation for idiosyncratic or pernickety decisions, SCB referee Jim McGregor is leniency personified. A case in point is heat 14 where Nathan Greaves runs into a fallen Kelsey Dugard. Under normal circumstances, Kelsey would be disqualified but, instead, takes part in the re-rerun Nathan Greaves looks likely to win until he suffers an engine failure. The race continues leaving him to gesture his chagrin from the centre green. Chelsea Lee-Amies is here with her sisters Shannon and Mercedes. It only seems like yesterday Chelsea told me she hoped to be the first lady speedway world champion. Nowadays she's left school and completed a beauty course but still finds work in her chosen area hard to come by. "It's Shannon and Mercedes who want to ride speedway and so does Stuart [her dad]."

Though Jordan can walk, sensibly he withdraws from the meeting. The sixth 250cc race of the afternoon (heat 17) sees Josh Bailey fall and remount three-quarters of a lap behind. With determination that typifies the attitude of all classes of this competition, he makes up considerable ground to finish third. Bryn Williams isn't happy with the raffle announcement, "Whoever drew the raffle tickets wants shooting, as neither myself nor the incident recorder won any of the prizes." There are many ex-riders in the pits this afternoon including Martin Dugard, Edward Kennett, Dave Jessup, George Barclay and Martin Hagon along with present-day riders Mark and Aaron Baseby. Linda Barclay drops by the canteen for a refreshing cup of tea. Apparently George didn't sign many photographs in his youth. So it's a surprise to see Linda proudly flourish an autographed photo of a rather handsome and dashing George attired in a West Ham race tabard (from the 1967 and 1968 season). At training tracks throughout the country, many riders benefit from George's guidance and experience. Sadly his enthusiasm as a teacher restricts his own track time. "George had his second ride of the season at Lydd yesterday. It's supposed to be part of doing the training schools that he gets to ride but he doesn't like to leave the boys, so he doesn't! They'd just watered the track before he was due to ride but he waited a bit so he didn't have so much cleaning to do afterwards. Did you see Len cancelled his meeting on Wednesday based on the forecasts for today (which has changed)? So, though they only had an outside chance with the cut-off date being today, they definitely won't be going there now! It does mean that Len won't have to run his free meeting to thank the fans. They probably wouldn't have anyway. [David] Mason didn't turn up at Stoke for whatever reason, I don't know."

The surprise package of the 500cc category, so far (this season), is Max Clegg. Despite missing round 1, he's surged up the cumulative score chart with a series of impressive performances. Heat 7 (aka race 21) requires two reruns before it finally runs to completion. Michael Neale and Sean Phillips find themselves disqualified to leave a match race between Adam Portwood and Max Clegg. Max looks likely to beat championship leader Adam Portwood until his engine failure. Though we can't hear what he says, Max is clearly angered. Kelsey Dugard wins the penultimate 250cc race of the afternoon. Bryn Williams likes what he's seen so far, "I hope I'm not being too premature but, if that young lad keeps up his rate of progress, he's going to be better than his dad! That should earn me a right hook." With Bryn's praise vaguely ringing in his ears, Kelsey wins his last race of the afternoon (four wins out of four) to ensure his automatic qualification for the final. Jack P. Blackburn looks likely to win heat 9 (aka race 23) right until his last-bend engine failure though he still almost manages to push home ahead of Patrick Crook who wins by the width of a wheel.

The 27th race of the afternoon (excluding classic demo heats) sees Luke Woodhull and Jack Collins vie for 125cc ascendency throughout. It's the kind of fast and competitive race that also typifies this afternoon's action as well as the comparative skill, dedication and enthusiasm of all the participants. Somerset mascot Henry ("the Wasp") Atkins comfortably wins the 125cc final ahead of Jack P. Blackburn. The rerun 250cc final is awarded – after a fall by Nathan Greaves – to Reece Downes. Afterwards Jordan Hazelden queries why Downes' first time out tapes touching offence didn't result in a 15 metre handicap? The 32nd competitive race of the afternoon sees – after a rerun – Max Clegg continue his meteoric rise up the cumulative table with a win ahead of Nathan Stoneman and Adam Portwood. An afternoon of bright sunshine is matched metaphorically by a wealth of future young British riding talent.

**11th September – British Youth Championship (Round 5)**

**Winners: Henry Atkins, Reece Downes and Max Clegg**

# CHAPTER 24

# Newcastle

## "We need more ideas and more young people. What speedway also needs is more fights!"

18th September

Floodlit

Because of the Great North Run (aka the Newcastle half-marathon) tinfoil litters the floor at Newcastle station and also the metro train out to Byker stadium. The sandy riders' car park at Newcastle Stadium is boggy and puddle strewn after torrential rain. With the sun obscured behind substantial dark cloud cover, the Diamonds curatorial team industriously blade lumps of shale from the track. With clipboard in hand, the gateman chats to referee Mick Bates by the doorway of the low-slung single storey building that contains the rider changing rooms, refreshment area and the Newcastle Diamonds Speedway Office. The second leg of the Premier League Knock-out Cup semi final versus Glasgow is due to get underway at eight o'clock, five and a half hours after the first leg at Ashfield Park. Barely interrupting his conversation with the referee, the gateman says, "George [English] isn't here yet. We lost by ten points 50-40." He resumes his conversation, "Jim McGregor was the ref so it was bound to take a long time. I've heard that Grieves and Newman got injured. No one can replace Newman around here but we're lucky because we've got Adam McKinna four miles away and sober." Mick Bates is on a mini Northern tour – Berwick last night and Newcastle tonight. The gateman asks, "What about Frampton last night?" Mick shrugs, "Sometimes you wonder did I get that right? But he wasn't even close! Why? "Because they've mentioned it on the Forums, I won't say what they've been calling you."

En route to the referee's box, Mick stops by my display table. "It was a shame last night at Berwick because there were a few incidents so we had to stop because of the curfew after heat 14, when Rye House were only 2 points behind. Some team managers would kick off but they were quite reasonable about it."
"Would they have won?"
"I don't think so because Berwick's top two were going well and would be hard to beat. They have to stick to the curfew strictly there and, who am I as the visiting referee, to say they should go on for a couple more minutes? Apparently the council is strict and they have a man in a nearby house who's on the phone to the council at 9.31! It's not the first time they've had to stop because of the

curfew after some incidents there. I was saying to my other half that this is one of the few tracks I'd pay to come to if this wasn't my job."
"Was it wet then?"
"No, a little bit like this but being banked it can handle the water a bit better than here. I don't get to come up here unless I ask the fixtures secretary because they have three refs in the area and I live at King's Lynn. It's interesting here but sometimes the track isn't quite right."
"Because of the greyhounds, they don't always have as long a time as they'd like to prepare on a Sunday."
"Ipswich is another one. Have you been there since they've changed the shape? They cut the corners but I don't know ..."

With the Old Firm (football) derby live on Sky, the first leg couldn't start until 2.30pm. Consequently, logistics as well as the distance between Glasgow and Newcastle dictate that riders, officials and fans alike need to put their feet down to cover the distance in timely fashion. Inevitably, riders being riders, they will race to see who gets to the Newcastle track first. With just under two hours to go before the scheduled (later) 8 o'clock start time, the pits remain eerily deserted. Two men look morosely on, apparently keen to will into existence the normality of the pre-meeting pits hubbub. "It's strange, nee body in the pits."
"It's strange at this time of night."
"It's probably carnage on the way down and some might not make it here."

Minutes later, Kyle Newman wins the riders' race down from Glasgow. With just over an hour before tapes up, dark clouds scud over. Joan English looks skywards, sniffs the air and looks out over the Brough Park speedway track. "The track was flooded at lunchtime after some really, really torrential rain. I didn't think we would be getting it on!"

With riders here, mechanics readying bikes and fans queuing the length of the fence close to the turnstile entrance, the curatorial team have shown skill in their work.

Over at the semi-divided hut that is one side the trackshop and the other the refreshment area (with a substantial range of confectionery), an avuncular man in generously sized black-and-white Newcastle Diamonds Wulfsport anorak chats amiably about the speedway world. "I tell you what, many people complain about that Plymouth promoter – what's his name – I don't think he'll be in the League next year but he pays us quicker than anyone else!" He asks me an often politically difficult question, namely: "What's your favourite track?" Reading is my standard and genuine answer. "It was ours too. They said there's gonna be another one but there's been nothing said since then. Who are my favourite promoters? Hum – Len Silver and Ian Thomas. Ronnie Russell, he's a great bloke! Well, except, if you disagree with him and then you think 'oh whoops!' He's better than his brother. I like Steve [Mallett] at Newport. They've done really well there. It's a great setup but a shame about the crowds. You shouldn't ever complain about them though. But his son, well? Because Steve sits next to me at meetings, he does as well. He's always texting. People used to complain about Jonathan Chapman but he is really very good. He's got lots of ideas and Buster just lets him get on with it. We need more like him in speedway! He winds people up because he's young. We need more ideas and more young people. What speedway also needs is more fights! When there's a good fight, the DVD sales go up. We've had a few good ones here over the years. Rob Grant, of course, oh, and that nutter Powell who threw his bike at Kenni Larsen, oh and Zetterström and Grieves after Christian got hurt." Sight of the track staff struggling to bring the blade out onto the track temporarily interrupts his flow, "What are they doing?" If 2010 was a triumph, then 2011 is more of a struggle for the Diamonds. "Last year we won everything but, this year, despite the injuries and that – it would still be nice to win a cup! You wouldn't want to go to Newport at the moment, I'd prefer Workington."

With the turnstiles about to open, Ray snatches a last-minute well-deserved comfort break ("I'll have a quick tab"). Very dark clouds now mass over Byker but my anxiety about rain is brushed aside. "I hope not. It's been rain all the way down to Hexham. What a rush back!" Club presenter Andrew Dalby's dedication to the Diamonds took him to Glasgow this afternoon rather than the Stadium of Light with his Sunderland season ticket. "It was 14-4 after three heats but, from heat 4 onwards, we got a few charitable donations from them so we're only ten points behind. It was a good meeting! 16 points would be good, 14 nice but I don't think 10 is enough!"

With the Glasgow Tigers Supporters' Club running a (rather luxurious) double-decker coach down to Newcastle, a substantial contingent of Tigers fans are expected. The Hamilton family are the first Glasgow fans I see come through the stadium turnstiles. Dad Colin updates me on the latest news. "James Grieves hurt his wrist and had ice on it all the way down and he's a track specialist here. Joe's got ignition problems – his first two races he stopped at the start line – so if he can get those sorted out he'll be good. We bought Alan [Dick] down in two and a half hours, so we've got it from the horse's mouth."
"You need him, he's the tactical genius."
"Err, no, that's Stewart [Dickson]!"

For a brief moment I think that I might actually sell a book at Newcastle for the first time since 2009, when Glasgow fan with an expectant look on his face rushes up and says, "Martin?" We're both disappointed. "I've a T-shirt for a man I've never met in my life who I've been told has glasses. Where's the trackshop?" Before the meeting starts, Barry Wallace pre-warns both teams about their opposition in the final. "I can tell you after nine heats at Workington, it's Workington 27, Newport 27. So that virtually puts Newport through to the final, especially after they won 59-31 at home." The first few yards of heat 1 sees Joe Screen immediately indicate that the Tigers are revved up for this cup encounter when, on a track that visiting riders regularly approach gingerly, Joe rides it aggressively. Newcastle presenter Andrew Dalby offers praise, "Joe wasn't hanging around on the M74 this afternoon either!" Newcastle need quick points on the board. Heat 2 sees Richie Worrall gate swiftly, closely followed by teammate Kyle Newman. Sadly for Diamonds fans, Kyle immediately slows to let both Glasgow riders through (though he soon reels back in Brough Park debutante David Bellego). Despite last lap advice bellowed from the grandstand ("Go on Kyle lad, hunt him down"), Kyle's unable to find a way past Theo Pijper. If there's pressure on the Newcastle riders to rack up quick points, then according to club announcer Barry Wallace, the CD operative faces similar intense pressure. "Bob's got 1,000 record requests to get through this evening, so here's the first of them ..."

While we wait for the riders for heat 3, Tigers fan Jim Fleming draws deeply on his cigarette. "We threw away a lot of points today. It'll be touch and go here, I think. I was chatting to a friendly Newcastle fan – I think him and his family were the only ones to come to Glasgow – but look how many we have here!" A substantial number of Glasgow fans cluster together in the vicinity of the start/finish line. With their distinctive accents and highly visible Glasgow Tigers Wulfsport jackets, they stand out. Unusually at a stadium where local fans are passionately audible, their partisan shouts for their riders is loud, continuous and unmissable. At number 3, Glasgow Australian wunderkind Nick Morris trails off at the rear of heat 3 unlikely to gain a point up until teammate Richard Sweetman (in second) suffers an engine failure on the last bend. Jim Fleming understatedly exclaims, "Oh, dear me" while other Tigers fans express themselves more strongly in the common language of men. Even before Sweetman's mechanical difficulty, Diamonds' reserve Richie Worrall (on a rider replacement ride for Joe Haines) looks the only likely winner.

Prior to the next race, announcer Barry Wallace conducts the initial lesson of his pronunciation master-class with Newcastle number 5 "Claze – Nay – der – mark". For the first lap Kyle Newman


26 Shades of Shale

and ex-Diamond James Grieves exchange first and second places until Grieves kicks on to show he still really knows his way round the Brough Park track (despite some surface roughness and extreme pressure). It's a drawn heat Barry Wallace greets with yet another pronunciation of the third placed "Knu – les Nedermark". Jim Fleming isn't surprised Theo Pijper comes last. "He's absolutely knackered! Pijper rode longtrack yesterday and was second for fourth overall and he was at Bordeaux airport at 4am this morning. Then he top scored this afternoon." With a very healthy number of smokers in the crowd, this is Jim Fleming's kind of club, "Aye, yes, that's right." News that Richie Worrall is to take another Joe Haines's rider replacement ride – this in heat 5 – prompts Jim to muse, "I don't think there's much wrong with Joe Haines, according to the boys from Newcastle. It's all very convenient!"

Joe Screen wins heat 5 imperiously though, slightly worryingly for Glasgow, Michal Rajkowski looks really out of sorts. Heat 6 enables Barry Wallace and Andrew Dalby to exchange their favourite pronunciations of David Bellego. Barry favours "Da – vide Belle – go" while Andrew prefers phonetics, "David Bel – lay –go". After an engine failure for Mark Lemon, Derek Sneddon wins heat 6 despite last lap pressure from James Grieves. Deek showcases his lap of honour celebratory wheelies serenaded by shouts of "Monarchs reject". Reading the race results, Barry Wallace pronounces "Da – vide Belly-go" like he's a suspect dish sent back during a disappointing holiday meal. Andrew Dalby provides some incisive analysis, "Every race, so far, has been won off gate three or four!" Nick Morris rides uncompromisingly and without fear to win heat 7. Kyle Newman applies pressure until he pirouettes impressively and falls. Andrew Dalby continues to reveal impressive pattern recognition, "That means the scores tonight have been identical to this afternoon [pause] for heats five, six and seven." The next race fascinates with an all-out duel between Richie Worrall and Theo Pijper, though Derek Sneddon wins to ensure a Newcastle heat advantage. A speedway fan first rather than solely a Glasgow fan, Jim Fleming confesses, "I like Derek Sneddon." Like many Glasgow fans, Jim travels widely to watch his beloved Tigers. "I was standing next to some bloke at Workington – I didn't know who he was but he was the Workington promoter – and he said to me, 'I'm going to sack James Wright' and he did!" James Grieves wins a drawn heat 9 to the joyous acclaim of the increasingly ecstatic Tigers faithful stood parallel to the start/finish line.

I'm impressed with the start marshalling. The gentleman concerned has an impressive range of unmistakably gestured instructions, plus he's also very firm with the riders whether they ride for Newcastle or Glasgow. Prior to heat 10, he gets Nick Morris to move slightly (Andrew Dalby notes, "gated three and a half there"). After a quick release of the tapes by SCB official Mick Bates, Morris fails to start well and is unable to make up ground on Richie Worrall.

Though it's not announced, we enjoy a mini interval prior to heat 11 with Newcastle six points ahead on the night and four points down on aggregate. The tractor driver magic wheels bend four with commendable intensity, endlessly circling in true Spirograph fashion. Barry Wallace uses the time to catch up on musical matters, "Another request – this time for Newcastle stalwart Joan English whose birthday is on Saturday 24th September from George, Sam, Pat, Adam and all the other friends at Newcastle speedway!" While bend three gets a comprehensive magic-wheel, Frank Sinatra's 'Fly me to the Moon' blasts out over the stadium speakers. Continuing a peculiar trend, Claes Nedermark is disqualified before the start of heat 11 under the two-minute time allowance. Not much escapes Andrew Dalby's attention, "Back to back exclusions for the riders in red!" Joe Screen then provides Kyle Newman with an unofficial further lesson in his speedway apprenticeship with a feint to the inside before an outside blast like the proverbial knife through butter. Though fans immediately mark up their official race-day magazine, Barry Wallace isn't able to formally confirm the situation, "I can't get the result at the moment because the referee is on the phone."

Collectively Glasgow fans are seized by hearty vocal optimism about tonight's performance and the season overall. Heat 12 sees Kyle Newman and Nick Morris enjoy a lusty first-bend battle where no quarter is asked or given. Morris overtakes stylishly on the third lap for a lead he fails to relinquish. The man ahead in the interval snack bar queue reviews the Diamonds performance so far. "It was old 'uns who've let us down. Lemon and Robson, oh, and Nedermark – you can't ask any more of Newman and Worrall! I must say Robson and Nedermark have been superb for us this season though." While we wait, some read temporary team manager Roy Clarke's account of his first ever visit to Plymouth. [1]

George English's "In Plain English" column also lives up to its name. This week George ranges widely covering Andrew Dalby's holiday in Switzerland, Diamonds injuries this season as well as the sad premature demise of Matej Ferjan, the end-of-season dinner dance and the Christmas party. Even more wonderfully, George chooses his words very carefully so the nuance and what's often left unsaid is often as relevant as the actual text. [2]

Joe Screen races to his fourth straight win and is followed home by James Grieves for a heat advantage (barring four non-finishers over the last two races) that ensures Glasgow reach the final. The Glasgow faithful react with a loud T-I-G-E-R-S chant. Tigers management are confident too, so much so they replace Theo Pijper with David Bellego, despite his solitary point (paid two) so far. Points aside, Bellego's been wholehearted and fearless with evident desire to succeed as well as zeal for the front. This continues in heat 14 where he gates, temporarily holds second and then really clatters the second bend safety fence. The initial silence of the crowd soon gives way to chatter when Bellego gets up. Jim Fleming previews the pending Scotland versus Wales Premier League KO Cup Final. "Did you see we raced Newport five times this season? They have a strong side but poor crowds!" Newcastle grab a win on the night through a maximum in the rerun of heat 14 and a drawn final race won by James Grieves. In addition to their aggregate win, one lucky lady (Glasgow) fan goes home £650 richer for the price of a £2 prize draw ticket. Tigers fans boisterously laud their waving riders onto the tarpaulin-covered dog track. They're rewarded with a series of interviews with Andrew Dalby. The first interviewee is James Grieves who responds to reluctantly to Andrew's forensic questions as if charged by the word.

"What a turnaround for you taken out of your third ride at Glasgow and yet you've had four wins here at Newcastle!"
"Yes."
"I think it's the first one of these [double headers] we've done in a decade, is it as gruelling as people say?"
"No, not really. It's not too bad."
"You must be pleased with your performance?"
"Yeh, I did alright, the track's really fantastic."

[1]"We eventually arrived at Boniface Arena (?) at 5.20pm, the man on the gate was optimistic that the light rain would stop in half an hour. Now the stadium is a bit of a culture shock, there's no cover for the supporters, just raised gravel banks around the track apart from the home straight which has a single decker bus coming off the fourth turn (this is for sponsors?) and a red and yellow box type structure for people to sit on. The bar is situated behind the first turn and is a large metal container, the same type that we use to store the tractors in. The pits were a bit on the small size, but the changing rooms were ok, as they were used by the local rugby team, whose pitch is situated behind the back straight. The track, is very small, there's very little straight (less than half the length of ours at Brough) and the bends are tight and narrow; you could possible fit it inside Brough. Because of the rain the track surface was very wet, but solid, apart from the fourth turn which seemed to be very grippy. There was a little bit of standing water, but this was removed prior the meeting commencing.

The starting tapes and finish line are not together, as the tapes had to be moved back at the start of the season, to allow more time for the riders before they hit the first turn. All the lads got there in good time and were in a very positive mood, not a single one of them even suggested cancelling the meeting, even though the Plymouth promotion were talking about cancelling the meeting until the next day, not that they said this to me."

[2] "Next Sunday, September 25th, there is no meeting here due to the Premier League Riders Championship at Sheffield. That meeting gets under way at 5pm and Mark [Lemon] will be our representative with Stuart [Robson] unfortunate to miss out as he wasn't in a PL 1-7 at the cut off point, no Management Committee favouritism there please note!"

"You've got to beat Newport in the final to win the KO Cup."
"That doesn't matter who is in the final."
"You've had some lively meeting in the past at Newport"
"Yeh, Newport is one of those sort of tracks."

With chatty Tigers team manager Stewart Dickson in front of his microphone, Andrew asks, "Stewart these guys just don't know when they're beaten." Stewart confirms the perspicuity of this remark and, when pressed upon the renaissance in James Grieves's form tonight (something Andrew makes sound akin to Jesus rising from the dead), Stewart remains phlegmatic. "Oh aye, he wasn't doing too good at all. I took him out of his third race because I knew he'd win his fourth race and he did and I told him he'd be the key rider here and he was!" After a career where his number of races exceeds his number of hot dinners, Joe Screen responds to "How well do you go round Newport?" with "How long is a piece of string? I don't know."

En route to the car park, long-time Diamonds fan George Grant – without whose kindness (in 1975) I wouldn't have the bug for speedway – once again now lives permanently in the North East (with wife Betty after returning from Spain). He regularly comes along to watch Newcastle race. Typically George takes a generous view of tonight's semi-final defeat, "With riders coming back from injury, two meetings in a day was just too much for us!"

**18th September Newcastle v Glasgow (Premier League KO Cup Second Leg) 47-43**

# CHAPTER 25

# Isle of Wight

## "I watch from the balcony but I can't watch the first corner"

20th September

Simmons family

Bad weather a fortnight ago saw the National League meeting between the Isle of Wight and the inelegantly named Scunthorpe/Sheffield postponed. Because of the cut-off date, in typical rich tapestry of speedway fashion, Scunthorpe/Sheffield (or is it Sheffield/Scunthorpe?) are due to meet Belle Vue in the play-off semi-final in early October. So, tonight's restaging fills a fixture obligation that remains – to all intents and purposes – going through the motions! Once they strap on their helmets and get out on the track, speedway riders being speedway riders will instantly forget this is notionally a "meaningless" meeting (and also grab the chance of some extra pay). Like they have for many other Islanders meetings this season, the three Amigos – aka Mike Corby (Islanders, Sittingbourne and Eastbourne programme editor), Alan Boniface and Ian Presslie – travel together from Brighton (via the mean streets of Worthing where they pick up Ian) to Smallbrook Stadium via the Portsmouth to Ryde ferry operated by valued long-term Isle of Wight Speedway sponsors, Wightlink.

Mike Corby's well-travelled (140,000 miles on the clock) royal blue Renault 'Harriet' smoothly eats up the 50 or so miles between Brighton and the ferry terminal. Ian reveals in Harriet's heritage, "You name a speedway track, Harriet's been there! We drove to the Czech Republic the day Mike got it new in 2000. Since then we've been to Croatia, Eastern Europe, Slovenia, Italy, Belgium, the Netherlands, Austria, Denmark and the Czech Republic. I might have forgotten some countries." Speedway fans the country over endlessly consult various meteorological information sources. Drizzle is forecast later on the Island but – given its microclimates – this might not include Smallbrook Stadium. Either way, it's two o'clock and we're on the road. Mike treats the possibility of postponement casually, "It's too late for me to get the call by now." Alan nods, "We don't want Bryn moaning on about it saying, 'They saw the forecast!'"

Alan and Mike are diehard Eastbourne fans. After a season of Bjarne Pedersen riding in almost talismanic fashion, the last two Eagles meetings have seen the Dane suffer a shock loss of form. Mike Corby takes the long view of his contribution and hopes that he'll return next season. "He certainly brings a bit more in the pits than Zagar."

[Jeff] "It was hard to bring less."
[Alan] "He only got 6 points from five rides last night."
[Mike] "Before anyone says that Bjarne is fiddling his averages – which he isn't into anyway – the Play-offs and the Knock-out Cup don't count!"
[Jeff] "Rolling averages prevent it."

When not analysing recent meetings, complex machinations or the ins and outs of the speedway world, conversation is random and wide-ranging. Alan confesses, "I've never been to a football match but, then, I haven't been to stockcars for three or four years. The only football match I've ever seen was a Speedway Eleven at Hellingly."
[Jeff] "Who played?"
[Alan] "I reckon it was 1971 or 1972 oh, God, I dunno – Barry Briggs, Reg Trott, Terry Betts and Jim Tebby. They call him 'Slim Jim' because he wasn't. He was a roly-poly man with no hair and big sideburns"
[Jeff] "Who did they play?"
[Alan] "A Celebrity Eleven – mixed celebrities you've never heard of like Ed 'Stewpot' Stewart and David Hamilton."
[Jeff] "What was the score?"
[Alan] "Goodness knows – 8-all or 13-12 or something."

Away from the track Alan is an avid memorabilia collector particularly smitten with Elvis, Marilyn Monroe ("I've cut down now") and Planet of the Apes. Like addicts the world over, Alan's quick to try to reassure, "Well, I don't buy so much now!" Sotto voce Mike Corby says, "Alan is a Jimmy Squibb fanatic too."
[Alan] "You probably never heard of him! He first rode speedway in 1947 and retired in 1976 at the age of 55. He was one of the speedway icons."
[Mike] "He rode with Ronnie Moore."
[Alan] "He had more or less a 100% record. He was always getting third for a 4-2 then bang they got rid of him!"
[Mike] "He wasn't a youngster."
[Alan] "He took some knocks at the end."
[Mike] "He went to Newport."
[Alan] "I always tell Neil Middleditch off – who spun round in front of him – for ending his career. Neil says, 'Oh that haunts me! People often say that to me. I'm so sorry what I did to that bloke!' It was a wet night at the last ever meeting at Rayleigh when they had eight men in a team. Ross Gilbertson didn't turn up so we only had seven. We was getting a 5-1 in heat 14 and Squibby went flying over the top of Middleditch and we lost the meeting. He got a broken nose, broken wrist and was badly bruised. It was quite a big track, which suited Jimmy because he didn't have to turn and he never ever rode a speedway bike again. That year – when we had the dinner dance at the Winter Gardens – he turned up with two black eyes and bandages. Whenever Dave Lanning used to interview him he used to say, 'You ain't getting rid of the old bugger yet!' The next year Lanning left and Bob Radford took over and said, 'I can't be having an old man in the team'. I haven't forgiven him for that! And then they got Jarman – but Pete Jarman always looked old. Squibby was 42 when he signed for New Cross. Guess what average he had?"
[Jeff] "I don't know?"
[Alan] "It was 11 in the Provincial League, what you'd call the Premier League now – and the only person ahead of him was Ivan Mauger! In his last season at the age of 55 he had a 5.5 average!"

It starts to drizzle as soon as Ian Presslie gets into the car. Immediately Harriet transforms into a mobile speedway badge auction room. Alan's brought a snazzy badges presentation board, "They're Eastbourne badges through the ages from the '70s to the present!" Possibly through

gritted teeth, Alan quickly laughs off my suggestion that Ian buy the lot for £10. On the subject of badges – albeit of the sewn variety – Ian's brought along the speedway version of Joseph's Coat of Many Colours. The dark sky-blue jacket is covered throughout in collectable or defunct club badges. Proudly holding it up Ian says, "That's history, that is!"
[Mike] "You stole that and I can tell you where you stole it from – Grindsted."
[Ian] "I bought it for a fiver from Mick Green."
[Mike] "Mark Greenburg?"
[Ian] "Mick Green in 1971 or 72."
[Jeff] "What club badges have you got?"
[Ian] "That one came from the first Birmingham track and I've been to three tracks in Birmingham! There's lots of defunct tracks – Teesside, Hackney, the old Belle Vue, Cradley, the first Birmingham, Reading, Barrow – the second track at Barrow, Park Road, Exeter, Boston, Canterbury, Coatbridge, Bristol, Eastville."
[Jeff] "California?"
[Ian] "I'm not that old! I've been to most tracks. Of all the current tracks, I've only got Leicester to do. Crayford, Ellesmere Port, White City, Wembley – I saw three world finals there including the 1981 when Bruce Pennell won it." Once Ian begins to list the continental badges, he quickly defeats my note-taking after the words "Lamothe Landerron".
[Jeff] "How do you spell that?"
[Ian] "It'll be on Google."

Familiar with each other's company, the three Amigos chat jumps from topic to topic including referees (Alan: "I don't know why Christina Turnbull didn't stop it" Mike: "From that angle she can't see"), litter ("the number of times you see one shoe by the side of the road"), recorded speedway race voiceovers (Mike: "I still think the odd ones you hear from Gareth Rogers are good"), obsessive collectors (Ian: "Alan collects germs and crisps – he's got more crisps in his bag than Walkers"), recommended speedway websites (Ian: "Aktuell speedway – a Swedish website – the results are always up there within a couple of hours") and surprising facts, (Mike: "It's funny that Geoff Boycott supports Man U – you'd expect him to support anything Yorkshire"). As we exit the A27 towards the fleshpots of Portsmouth, Ian ponders our crossing, "I wonder who's on the ferry this week."
[Jeff] "Who was on last week?"
[Ian] "Rick Wakeman."
[Jeff] "Did you ask him about his famous concert on Ice at Wembley Arena?"
[Ian] "I haven't heard of it. I said to him I love 'Grumpy Old Men' so I asked him if he was going to the Island for comedy or music. It was music in Sandown. I asked him if he still lived in the Isle of Man and he said he'd been in Norfolk for 11 years."
[Mike] "Jethro was on the ferry a few years ago."
[Jeff] "What Jethro Tull?"
[Mike] "I said that! He's a West Country comedian."
[Ian] "Cornish."
[Mike] "That's West Country."
[Ian] "It's Cornish, they're very particular!"

As we cross Portsmouth city outskirts, Alan remarks, "Here we are again."
[Mike] "Hopefully for the last time this year."
[Ian] "I admire Tony Clarke."
[Alan] "He used to ride for Wembley and West Ham."
[Ian] "For coughing in Backtrack for stealing those [Russian] bikes."
[Alan] "He looks a funny fucker."

Unlike the summer holiday high season, very few vehicles wait for the signal to drive onto the four o'clock ferry. There are a couple of speedway riders' vans including that of Islanders captain Nick Simmons (who's come along with wife Zoe and 14-month-old daughter Summer). SCB referee Chris Durno waits too. To pass the time, he rootles in his boot rearranging special priced litre bottles of Tango. Last night's Elite League speedway Play-off semi-final first legs didn't light Chris's fire. "The speedway just didn't get going last night and both away teams were gifted their way back into it. Even the crowds weren't what they used to be!" Chris likes to visit every British speedway track each year. "I hadn't been here until a couple of weeks ago – I came as a passenger with Dave Robinson – and now I bloody get allocated it to force me to return again this week!" The referee's bush telegraph communicates at warp speed. "Did you hear about the last Plymouth meeting? Christina Turnbull was there. They had 40-foot flames after the pallets stored behind the track caught fire and, when the Fire Brigade came, they blocked the service road and kept everyone parked inside the stadium there until midnight! Who knows if Plymouth will be in the Premier League next year? People say they're very bad for other promoters' businesses because they get poor crowds, huge scores with big bills and huge travel costs. There won't be many there tonight. The refs who've been over to the Island in the summer have said the crowds have been good but, tonight, it'll be 100 or 150 people."

Once we're on board and round the table with refreshments, Mike asks, "What's the best gig you've ever been to?"
[Ian] "Glen Campbell at the Guildhall in Portsmouth."
[Mike] "My best ever rock band – I'll probably have a different answer on a different day – but, today, let's go for something predictable: Led Zeppelin! My best ever gig would be the original Fleetwood Mac with Peter Green at the Assembly Hall in Worthing in 1968."
[Ian] "They were all there that year – Jimmy James and the Vagabonds, Geno Washington and the Ram Jam Band, The Who!"
[Alan] "My favourite band would be nobody but my favourite solo artist is Elvis Presley and my best ever gig was at Wembley Arena in 2000."
[Mike] "He wasn't there!"
[Alan] "They did the live Elvis concert. They brought all the original band. It was brilliant and that's still touring today."
[Ian] "Another good gig was the Bootleg Beatles."

My attempt to kick off an alternative discussion about what's the most embarrassing gig you've seen prompts Alan to declare, "I couldn't say. I saw Glen Campbell but didn't like him. I liked his backing singer Diana Solomon, she was flippin' crumpet."
[Mike] "I don't know his name – a white American blues singer at Hailsham in a beautifully restored theatre – but he didn't do too well."
[Ian] "He was as pissed as a rat."
[Mike] "He didn't know what he was doing."
[Ian] "It wasn't long before they took him off. Manfred Mann and the Blues Band they were excellent too."

It's not uncommon for riders to have pre-meeting and pre-race superstitions or habits. Sat in a booth together, Nick Simmons and his wife Zoe have their own rituals on the way to Smallbrook Stadium too. Zoe reveals, "We have our superstitions. We have to have a caramel shortbread and a packet of Quavers every trip and some cheesy chips when we get there!" 14-month-old Summer ("She was born in July but named in January") is a curious, smiley baby ("She's good with men"). Islanders captain Nick enjoyed a good campaign this season. Tonight is the last official fixture, "You just want to finish the season on a high so five races five wins would be good! After a few weeks, you've forgotten all about it though." Though technically not a superstition, Zoe is a creature of habit. "I watch from the balcony but I can't watch the first corner. By the time he's on the home straight, I can open my eyes." My polite reply ("When he's leading") stirs Nick, "Hopefully! It's been a good season so it would be good to carry it on!"

Referee Chris Durno joins the three Amigos. The consensus is they'd prefer the sport didn't rip itself apart (again) this winter. There is also unanimity over the likely triumph of the Pirates in the Elite League, "It looks like Matt Ford will win it so that should keep him happy!" Mike, Ian and Chris are all obsessive international speedway travellers. Chris sheepishly admits, "I've broken into over 200 tracks now. The more tracks you do the harder it gets. It was weird my first FIM meeting [as a referee] – Tayac in the Dordogne Valley – was my two hundredth. Recently I picked off Munkebo – just above Odense – we went there on a Friday." According to Ian, such expeditionary trips can have their frustrations, "We went to Romania, but when we got there, there wasn't enough riders so they called it off!" Nick Simmons interrupts talk of obscure international speedway tracks to ask. "Who won the longtrack?"
[Alan] "Joonas Kylmäkorpi."
[Nick] "I wanted to know who won the 350s."
[Alan] "Oh, I dunno, we've let you down."

En route to the car deck, Chris continues to relate some highlights of his travels to foreign fields that are forever speedway. "I said to Tony [Steele] I'd like to do [as FIM referee] the ice speedway in Siberia. I go every year anyway and he said, 'Join the back of the queue'. A few years ago no one wanted to do it, but it's a different discipline so it appeals." Chris still enjoys the challenge of his British speedway duties. "Newport versus Workington was one of those meetings when you have to make difficult decisions. It was dry on the inside and morass on the outside and I kept saying to [Steve] Mallett 'What can you do?' They really worked on it. The hardest decision as a ref is saying 'it's okay and you're gonna ride'. When they go out for heat 1, your heart is in your mouth in case that's not the case! Luckily it went ahead because the last few races were cracking." Chris is astonished at my news (from the *Speedway Star*) that FIM referee Frank Ziegler is to retire. "I'm amazed. He's a knowledgeable passionate guy. He is Mr UEM. He's almost like a one-man manager for strange tracks and places! He gets a bad press but, the first time I met him, he went to the top of the tree for me. His two sons are riding now so, maybe, it's a conflict of interest?"

Inside the referee-cum-announcer's box that overlooks the Smallbrook Stadium home straight, it's a surprise to see Bryn Williams preparing for meeting presenter duties since (after many seasons) last year Bryn was replaced. Tonight he's here because Tim Allan (aka the speedway presenter formally known as Tim Helm) is away. "I've cut down this year. Today is my 111th meeting and, by this time last year [consults diary], I'd done 127. Saturday I'm at Rye House, Sunday I'm at King's Lynn for the Southern Riders' Championship, Monday Eastbourne, on Tuesday I'm off, Wednesday Belle Vue, Thursday I'm off and Arena Friday. Have you seen this?" Bryn waves a leaflet tucked inside the official match-day programme. Isle of Wight speedway club clearly sweat their assets nowadays since – on Sunday 23rd October – they hold the "2011 Autumn Smallbrook Wedding & Party Fair" "'Ere, look, I misread that, I thought it said penis not genius!" (It actually said Gems!) As usual, Bryn came over with Islanders incident recorder Chris Golding, "The quack diagnosed him as diabetic and he's been told to 'eat little and often' but he's forgotten about the 'little'!" Bryn has no patience with speedway meeting presenters with ideas above their station who come to believe that they are, in fact, the centre of attraction, "The only people who matter are those with their bums on the saddles, we're just here to help the show along – that's the way I look at it."

On the steps outside the box, there's a mini tableau that typifies speedway a couple of hours prior to tapes up. Chris Durno asks, "What do you think of it?" Graham Trollope glances trackwards, "Nice and wet! To be fair, they've always got excited with the water here." Out on the track some

riders conduct a closer inspection of the shale. Bryn shrugs, "They're all digging their toes in. They ain't got a clue what they're doing. Danny Warwick's raked it all and now they're digging it up again. Barry Thomas would turn up, look at the track, make a roll up and go off and cut the tyres. They don't even have to do that nowadays. Are you going on the track walk, Graham?"
[Graham] "Not bloody likely. I've seen too many already."
[Bryn] "You usually do it on a tractor."
[Graham] "It's nice to watch someone else do it. I went into Sheffield at seven this morning it started raining so, I thought, sod it – day off! Trouble is you have to work round the dogs. I could do without the cars at Scunthorpe and then I have to work through the night for the PLRC."
[Rider with broad Yorkshire accent] "You have to make the call when you can because you can't get a bloody signal here, I hate the place!" Chris Durno returns from his solo track walk inspection of the safety fences, kickboards and start gate area (including flags). "There's a massive ridge on the racing line [bend one] where the tractor's been that they'll have to get rid of." Another man arrives to tell Graham, "Go and have a look at that berm on the first corner!" Graham, "He's said about it."
[Chris] "'Ere, Bryn, has Nick Simmons got the bronze helmet?"
[Bryn] "I dunno."
[Chris] "You work here, it says that in the programme. That wasn't a whopping crowd at Lakeside last night?"
[Bryn] "It was good but it was on the telly. Is it Scunthorpe or Sheffield?"
[Chris] "I call it Sheffield – a big city town! When's the last meeting?
[Bryn] "It's this one."
[Chris] "What fireworks and riders' speeches?"
[Bryn] "I dunno?"
[Chris] "How have they done this year?"
[Bryn] "Okay but not enough."
[Jeff] "Are Lakeside gonna win at Eastbourne?"
[Bryn] "I dunno, the season's gone on too long for me. They didn't expect to be there anyway!"

Club co-promoter and co-team manager Kevin Shepherd is on the first bend in deep discussion with two members of the track staff. An approachable and polite man, Kevin breaks off to say hello. I tell him my latest book *Bouquet of Shale* is co-dedicated to his mum, the late Eileen Shepherd. "If you mention it to Dad it might upset him because today would have been their 59th wedding anniversary! Like all people, some days he's good and some days he's not. I can't tell and, when I spoke to him this morning, he was alright. He'll definitely be here later."

Islanders Environmental Officer and Pits Marshal Jane Morris has worked in the pits for about a decade. The discoloration of her high-vis yellow fluorescent jacket signals her length of service. "The away riders are all getting changed but our riders get changed in their vans. It's a sore point. They're here to put on a show and walking up and down is all part of it. The fans see them, the kids get the bug and they find sponsors! No one has been up to the bar for a drink after – they shoot straight off in their vans. They weren't even going round for a lap afterwards. Alright, if they have two on the trot, they can't but Pav used to stand at the pits gate and wave them round and tell them to put on a show! About halfway into the season I got sick of it and I went down and told them 'there'll be an Island flag for you on bend four so go round!' They said 'okay' and have since. I've been doing this 10 years now and you have to be firm or they run all over you! I've been kicked, spat on and sworn at but when the racing's on you let them get on with it but, then, have a word afterwards. The other week a 17-year-old visiting rider rode down the pit lane very quickly and nearly knocked over a mechanic. I went down to see him and he told me to eff off. I spoke to his manager and he told me I should have a word. When he came in next, he kicked a hole in a bin so I went to see him again and he told me to eff off. So I told him I was a SCB official so if he'd said it as

I'd walked away I'd ignore it but he'd said it to my face. He didn't think it serious until I went to the referee and she said 'send him here!' When I got back his captain said 'what if I go instead?' I told him that was fine but if he got the fine that was his lookout. The rider was sat on his bike saying 'I've got to ride.' I said 'the ref is waiting and won't start the race without you'. She told him he had to apologise or get a fine. He came back and just about said sorry. The riders know I'm fair. I'm not biased to our riders. From 7pm onwards, they know I don't speak to them unless I have to. I might have a word afterwards. Five laps Les isn't here tonight so there's no flag. It's from when Fat Rat was here. He made them but they've run out now."

By the raffle table close to the entrance turnstiles, staff carefully study the photocopied letter from Brian Ginders (Chairman to the Board of Directors on behalf of Island Speedway (IOW) Ltd) about a possible staff meeting at Smallbrook Stadium next Tuesday night. [1]

Michelle and Sue ("Sexy Sue" with "the voice of a Town Crier" according to Bryn Williams) clutch their letters while a gentleman suggests his blueprint for the future, "Start at Whitsun Bank Holiday and run for 14 weeks." Sue's delighted at the staff consultation, "I don't have any grievances but I have a lot of ideas!" Clerk of the Course Graham Arnold arrives holding an attaché case and the hand of his partner. "I think they should start earlier but they say they can't because the planning is 7-9.30 or 10 if an ambulance goes on to the track. They need to change the planning! I bet if they said they'd start earlier and finish earlier, the council would be fine about it! People [holidaymakers] are already on the Island so, by six, they want something to do. They're not in the Premier League now so they have to think about how they run their meetings. They say they run twice a week in the summer, but they don't! It's speedway one meeting and something else another. They should do the odd Saturday or Sunday like Malcolm does at Lydd. Okay Wightlink don't like that but the odd one would make a difference in summer. When the half terms were on, they didn't run meeting then but they're running now! For Dudley the other week, they had a very small crowd – much less than 100. If the meetings finish by nine o'clock, the riders might go to the bar? In the National League, they all have to work the next day so they have to rush off for the ten o'clock ferry."

Graham heads to the pits followed, moments later, by a gaggle of Sheffield/Scunthorpe riders moaning as they pass, "They still ain't took that hump off!" Sue's not quite sure what's going on either, "Why is everyone coming down here to get dressed tonight?" Sue believes the consultation should be wide ranging, "We should ask the fans too! We need to do it while the ideas are still hot! To be fair to him, Brian is really good. There's basic things that could be done straightaway. When you drive past here it looks as tatty as anything. It needs painting up and someone artistic painting a rider or a flag or something? Alright we know what the locals are like, they don't come even when we have the Sky meeting and it's free. Well, they did come but didn't come back! So we need to promote it to the visitors. Okay, at the start of the season, we had the Press and Practice and put up loads of posters. But, by the time the season starts, they've taken them all down so no visitors get to see them! There's so many fairs and events we should take a rider along with a bike or something. Okay, it costs money but many people don't know we run! When I used to come over with my girls on the bus in my Islanders jacket people would ask about the speedway. When the Buxton fans came they said no one knew about it in West Wight!"

In his trademark predominantly red Islanders large-sized Wulfsport anorak, Albert Shepherd arrives with carrier bags packed with vital refreshments. "Hello Steve, nice to see you! It would

[1] "Dear Team, We have looked at things this year and decided that we should hold a meeting of all staff at the end of season. The idea is so that you get your say on how to improve things not just for you and the job you do but we can look at where we can improve for the future of the club etc. We are thinking of next TUESDAY EVENING at 6.30 till around 8.30pm here at Smallbrook. Please put your name at the bottom of this letter and hand it into the tearoom tonight so we will know how many is coming and it would help if you put down what position you fill at speedway. You are strongly encouraged by all the management team to come along and contribute, all points of view will be listened to and if viable acted upon. I would like to take this opportunity to thank you all for your hard work and commitment you shown to this club. Yours truly, Brian Ginders, Chairman to the Board of Directors.

have been our 59th anniversary today. I've still got her ashes in the house. I can't bear to part with them. I still can't touch any of her stuff – all the jewellery. There's four wardrobes full of clothes and ten pairs of shoes too." Another man passes in an almost identical Wulfsport anorak. Once he's out of earshot Albert whispers, "That's Brian Ginders. Without him, we'd really be struggling. He owns lots of Nursing Homes and we all know how much money they take. It's been a mediocre season but we're still going which is the main thing!"

With the meeting about to get underway, Bryn Williams quickly runs through the introductions without a rider parade. He starts with the visitors. "Four is Benji Compton – one of the rising stars of British speedway – as they all are!" Bryn also has a warning. "Jeff Scott is over by the entrance in his shorts. He's a renowned author it says on my piece of paper. He certainly wants to beat last year's sales figures of zero. It says here he's got a Tupperware container and so has the incident recorder. Our incident recorder, Chris Golding, is very observant when it comes to food but also with other things including programmes. The Islanders team managers are Chris Hunt and Kevin Shepherd but not Gary Cottham who's got more important things to do like ride at number 4 this evening! We've dispensed with the rider parade because of the threat of rain. You can cheer!"

To signal the racing is about to get underway, over the tannoy system, speedway music par excellence – Deep Purple's 'Smoke on the Water' – briefly blares out before it's paused. Alone on the raffle stall, Kay tells me, "There's no longer a double-decker bus service from the ferry to the stadium. There's no bus anymore – there's taxis supplied by Wightlink." Kay also updates me on infrastructure developments within the stadium, "There's some new toilets by the pits – they're brick built!"

The Islanders open with a 5-1 through Warwick and Simmons. A drawn heat 2 follows. Blue-helmeted Gary Cottham gets the tapes caught on his helmet visor so starts 15 metres back for his troubles. A Steve Worrall engine failure gifts Cottham a point he nearly doubles with a last-gasp dash for the line. Before heat 4 starts, we enjoy a brief break for track watering to combat the billowing dust noticeable throughout heat 3. By the time Danny Warwick wins heat 6 (Albert: "I think Danny's got a lot to prove, he's had a few bad meetings"), the Islanders comfortably lead until Sheffield use Worrall as a tactical. It transforms the meeting – with Benji Compton second – the 1-8 narrows the scoreline to 24-21. Danny Warwick continues his imperious early meeting form with yet another race win in heat 8 to make it three from three. Prior to heat 9, we enjoy yet another track grade. Whatever they do to the shale certainly suits the Sheffield riders since they provide six heat winners over the remaining seven races. Lee Smethills's heat 11 victory (followed home by Lewis Kerr) gives the visitors their second maximum heat advantage and a narrow 1-point lead. As the drizzle starts, weatherman Bryn Williams notes, "The rain has arrived and so has heat 12." They cancel the interval and crack on with the superfast finger of SCB official Chris Durno prodding the buttons.

After his first race engine failure, it's clear Worrall relishes the 385-metre banked circuit of Smallbrook Stadium. He zooms home in heat 12 at impressive speed – looking all the better under floodlights haloed by darkness – to win his third successive race. Sheffield's 3-point lead evaporates in a rerun heat 13 that, first time out, sees free flowingly fast Ashley Birks crash and get disqualified. Even without the substantial number of young children from Swindon, tonight's crowd – to my untutored eye – looks a good one. With two races to go, it must be close to bedtime or perhaps they've seen enough? Whatever the reason, the children leave the stadium. Up in the box Uncle Bryn Williams says, "Bye kids!" It's a message they greet with enthusiastic waves, "Bye kids wherever you're going back to?" The experience and strength in depth that's seen Scunthorpe/Sheffield top the National League table this season reasserts itself in heat 14 (Benji Compton wins with reserve replacement Adam Wrathall second). The ungainly 42-45 scoreline prompts Bryn to say, "Well, it's a topsy-turvy match."

Drawing deeply on his cigarette while the riders dig for victory at the start line in heavy drizzle prior to the start of heat 15, Chairman Brian Ginders looks confidently to the future. "They're getting back on track and we're making changes. The clubhouse has a new boiler, new furniture and we have a brand new toilet block: ladies, gents and disabled. We're going to make other changes but you need money coming in though! We have to run it as a business! Next year, we're going to work at getting people back because we've lost a lot. When you come next year, you'll see the difference in everything and the crowd!" To win the meeting the Islanders require a last heat 5-1 but Birks and Worrall immediately extinguish this possibility.

With the drizzle turning to rain, Chris Durno runs 15 heats (and two reruns) in just over 90 minutes. It's the one speedway location where, if you finish early, there's no benefit since you still have to wait for the ten o'clock ferry! In the ferry queue, Alan Boniface accentuates the positive. "We're in the warm, we're dry, the meeting went ahead and no one got hurt."
[Mike] "One or two Premier League riders make a big difference at this level."
[Alan] "Trouble was, we were carrying too many passengers! And Gary Cottham, who's good for 6 or 7 or 8 every meeting, only got 2 after he got excluded in his first race."

While we wait, talk turns to the shape and topology of the new Leicester track ("The best racing is on tracks with short straights and wide sweeping bends not long straights and tight bends. It's very strange to do that when you've had a blank canvass!") and Mike Bennett, ("I can tell you where Mike Bennett started – Boston – we gave him his first speedway job. I even praised his programme column"). Mike Corby has a difficult question for Miss Marple (or Hercule Poirot), "I was driving through Hassocks the other day – because I do my shopping in Burgess Hill – and I saw Jason King's van parked outside the new – well it's been there two or three years – pet crematorium they have there. When I saw Rob – father of Jason and Danny – at Arlington I said, 'What were you doing at the Pet crematorium?' [2] and he said, 'I haven't been there but, funny enough, I am driving Jason's van at the moment because I have to drive him around because he's injured'. I suppose it will have to remain a mystery forever!"

### 18th September Isle of Wight v Scunthorpe/Sheffield (National League) 43-50

[2] Mick would subsequently tell me, "I solved the mystery! You know you see something new spring up and you think 'it's a pet crematorium?' Well, it's not – it's a human burial ground! It's non-religious (humanist) in woodland called Clayton Wood. Funny enough, I'm going to a service there on Monday.

# CHAPTER 26

# Sheffield

## "He's been doing that all season – turning left!"

29th September

Replica top model

Like the rest of England, Sheffield sizzles in unseasonable almost record-breaking sunshine. Inside the warm glass-doored Sheffield Speedway Office at Owlerton Stadium, Neil Machin's on the phone to co-promoter David Hoggart. "What the fuck does he want a collar and tie for? He's only here to sell books!" Phone call over, Neil resumes alternately shuffling the paperwork and stacking cashboxes behind the counter that separates arrivals from the inner sanctum. "I'm just wading through things and trying to earn a crust. Selling anything anywhere's bad! The world's struggling! I'm trying to wade through things without any help from Julie, who's on holiday. Did you want a tea or coffee? In another hour, there'll be no chance of a cup of tea!"

Photographer Andy Garner is about to head off to the centre green with his camera and a framed Shane Parker presentational photograph. "I must show you this. I took some pictures for Hugh Skidmore and he said 'I'll sign it for you'. I'm not being funny, it looks like a 5-year-old has done it!" While he waits for rescue (aka Andrew's arrival), Neil continues to agitate his sheaf of envelopes on the table and expresses scepticism at the unsubstantiated rumour of Keith Denham's heart attack. Luckily Workington Comets team manager Tony Jackson will soon be here to quiz. My opinion "Tony's a nice bloke", gets rebuffed, "I don't really know – I don't have to live with these guys you know!" Also compulsively fiddling with his cashboxes, Neil talks to himself, "So I've got the turnstiles and cash floats." Andrew arrives and Neil runs through the envelope recipients, "I've got Trollope, one of the paramedics, Lou, John W, Farmer John ..." It's an endless list. "The riders' wages are all in here – I've left them open in case something has to go in – I've done copies of the pay sheets that are on here. There's a form somewhere that has to go with Doodson's cheque. I had to check back through the cheques to see how much we paid them last time."

After a £10 million refit, Owlerton Stadium's Panorama Room justifiably gains plaudits for its aspect as well as the quality of the service and menu. Restaurant manager, Craig, pops in for an update prior to getting on with his work, "Have we taken any bookings?"
[Neil] "I haven't had any bookings! So I don't know if that's good or bad."

[Craig] "Are you having your usual table?"
[Neil] "Eric and Winston are coming but at half eight so I'll settle up with you then if that's alright?"
[Craig] "That's okay."
[Neil] "I haven't had a single call, it's a fucking worry? How many did you do on Sunday?"
[Craig] "170."
[Neil] "I thought it was 200."
[Craig] "Some didn't turn up."
[Neil] "I always wanna do well on that [PLRC] meeting because it's expensive for me and for you."
[Craig] "Aye it were fine. I had one stroppy woman. She wanted to walk out without paying."
[Neil] "What the fuck for?"
[Craig] "She told me she came regular but she don't because I'm there every week."

Half of Sheffield's speedway's regular presentation team, Shaun Leigh listens with feet up while Neil recounts his busy week. "The PLRC all day, Sunday I was here, Monday I was at Glasgow, Tuesday I was here and Wednesday I was at fucking Rugby – I'm here now and then we've got Plymouth tomorrow – so it's even more money." Neil jointly runs the Sheffield/Scunthorpe National League team with Rob Godfrey. After the first leg yesterday (away at Belle Vue), Sheffield plan to stage the second leg of the Play-off semi-final next Thursday If they go through, in the final they'll meet either Mildenhall or Stoke. Neil and Rob tossed a coin to see who stages what (semi-final or final) where (at Owlerton Stadium or the Eddie Wright Raceway). "We'll give Rob Godfrey another fucking big pay day. If you can't get a crowd for a final when can you?"
[Shaun] "I dunno?"
[Neil] "He can't get crowds but then he has a different cost base than us."

Cup competition conversation is cut short by traffic news delivered from the pits by Sheffield captain Shane Parker. "Emiliano's called in to say there's a big smash on the M1 and that he hasn't moved for half an hour."
[Neil] "What the fuck's he doing on the M1?"
[Shaun] "He'll be coming across."

All in this office are power users of the British road transport system so – after a brief discussion – it's quickly established that Emiliano is the only Sheffield rider likely to be affected apart from, possibly, Simon Lambert. Without really leaving the doorway, Shane Parker reaches over the office counter to retrieve his large bottle of water from the fridge. Neil asks, "Do you want a tea?"
[Shane] "I don't drink tea."
[Neil] "How about a beer then?"
[Shane] "I've never touched a beer in my life."
[Neil] "Bourbon and coke?"
[Shane] "Tell you what, I bought a double in the bar here for £6.20. In me local, it's £4.65 for a double and they wonder why they don't get any business? It's £20 at the supermarket!"

Neil ostentatiously tucks his smart white shirt deeper into his trousers. "I'm just moving off a 32" waist and I was that when I was 17! Canderel's the secret. It's such a discovery. Ten years of sugar puts on the pounds. You should think about it?"
[Shaun] "I'm okay."
[Neil] "How old are you?"
[Shaun] "45."
[Neil] "45 isn't young! Me mum's 93 and she still thinks I'm a kid and I'm 63. I've a brother who's in his 70s and she thinks he's a kid too! I wrote to Canderel wanting to do a promotion and they never even fucking got back!" Neil doesn't share my amusement at the notion Canderel and speedway aren't natural publicity bedfellows. He changes the subject. "I got a bollocking on Sunday off

Gilbert because I didn't have a risk assessment to set fireworks off."
[Jeff] "That sounds a bit excessive."
[Neil] "There were 96 people killed down the road."
[Jeff] "You tell me that every time I come here."
[Shaun] "It's very strict here. If you look at the back straight here – and you look at the back straight at Belle Vue or Stoke – and we can't have ours open! You know that our council is so strict it's scary! It's a good job I didn't bring the flame lights – Gilbert would have gone balmy. It would have been funny if he'd asked, 'where is my risk assessment?' and I could point out its silk flames. They just look real from 100 yards!"

With practised ease Neil takes a cigarette from his (red) Marlboro packet, lights up and inhales deeply, just as tonight's SCB official, Craig Ackroyd arrives in the doorway. Last weekend, Craig officiated at the Speedway Grand Prix round at Gorican in Croatia. This step onto the bigger stage is good news for Craig's career but, away from the track, there's even cause for greater celebration since Craig's wife is pregnant with their first baby ("It's due on March 8th right on time for the start of the speedway season. I don't know if I'll be too tired to do it or doing it to get a rest!"). Whether it's a boy or a girl, Craig's likely to pass on the speedway officiating gene he inherited from his father Paul. Prior to any speedway meeting, each SCB official checks the previous ref's meeting report for any outstanding housekeeping issues with the track, facilities or equipment. Craig waves the report about for effect but doesn't really look at it, "There's never anything on here." Neil shrugs, "Actually Graham Flint were here on Sunday so there could be! Actually he said the ref's box was messy."

Before anyone can get the Hoover out, an irate Ricky Ashworth looms in the speedway office doorway and needs no second invitation to show off substantial dark purple bruising at the top of his arm (after his fall at the PLRC last Sunday). "It's more than a broken fingernail like Michael Max said. I'll fucking have a word next time I see him."
[Neil] "It could be worse, it could be my knob."
[Craig] "How was Glasgow?"
[Neil] "They've screwed up, they've got six fixtures and only four Sundays left. They started late and then they had fucking sun-offs."
[Craig] "Dry-offs?"
[Neil] "That guy at Plymouth is just as bad. He had sidecars last Friday and there's five meetings to go still."
[Craig] "He's the sort that wouldn't mind not finishing."
[Neil] "The fines will be massive, so he will care. What he's doing is running a cheap and cheerful team at home and not even booking guests away!"
Tigers team manager Eric Boocock hovers, Andrew tells him, "He's stuck on the motorway."
[Eric] "I know, looks like we're having a parade at half past nine!"
[Shaun] "Where are they stuck?"
[Eric] "M18 junction."
[Neil] "Who else is stuck? Lambert?"
[Eric] "We don't know where Lambo is?"
[Shaun] "Shane said he was stuck too."

Eric chews the fat in staccato fashion about Glasgow ("the track were fantastic – he's a trapper and it was smooth at the start and really really fast!"), rider absences ("we saw his van outside the Casino at 11 so he knew he weren't going"), riders changing tuner because after they'd paid a £1,300 bill the tuner ignored them the next time they saw them ("I said to him your Jawa engine is completely genuine, it won't have any funny parts – fake rods, fake screws, fake anything! So he was stupid! How come Shane's been there eight years and won about 16 world titles? So more fool you!") and why during a speedway meeting isn't always the best time to try and chat with the riders, management or officials ("when you're at a meeting, you've got your own thoughts so you walk past everyone! Often they say to me 'I spoke to you at the meeting' and I don't even know they have!")

Round in the shade at the trackshop, all the merchandise is attractively laid out. Jeff Dooley enjoys a well-earned rest along with Eleanor Ford (who studies psychology at Sheffield City College). Jeff Dooley models a Sheffield speedway replica top. He's understandably and justifiably proud of his handiwork. "It's not ironed on but proper done. I looked all over the world. I looked on the internet and eventually found someone who made ice-hockey shirts. He's literally five minutes from where I live. It's fantastic quality! It's been a godsend for what we do." Sheffield speedway umbrellas (rather than replica tops) were more relevant at last Sunday's PLRC, "The rain spoilt it". Speedway fan Bob isn't so sure, "It weren't rain it were riders! Let's talk about Lemon and Gjedde." Jeff Dooley prefers a meteorological explanation. "It were like this then black clouds came over at quarter past one, drizzled for an hour and the track were a mess. We had a good crowd and that bit of drizzle spoilt it. If it gets a bit deep and wet on the outside, some won't even go out there! This season has not been bad. Crowds have held up well considering the climate. I don't think you could ask for more. People are a little bit more choosey. They've just taken a phone call in the office tonight, 'Who is it tonight?' which tells you they're not just coming along but finding out who it is. This season it's been excellent to watch Shane Parker – it's been pure entertainment – and a good local lad, Ashley Birks, who's come into the side. He's a nice kid not a gimme. There's a local hospice – Bluebell Wood – that Julie Reading supports. Riders go there for the afternoon and take the bikes and chat to everyone there. Paul Cooper and Josh Auty. Before he went there, Ashley came round for some photos to give away to the kids. For a 19-year-old lad to be thinking of others like that tells you a lot about him!"

Many big name authors come to Owlerton. "Ivan did well again, I thought he'd struggle. Shawn Moran did well. There was an excellent night the night before. 'An Evening with Shawn Moran' with Seth Bennett interviewing him. He's still got a fantastic attitude to speedway. He still wants to put something back into the sport now he's on the straight and narrow and lives in Germany. He wants to help British speedway. He's seen all the pitfalls so can tell them what to do and what not to do!" Jeff appreciates the dedicated commitment of Sheffield Tigers fans going to Rye House and Glasgow. "That's dedication to their speedway. Some of them will be at Plymouth tomorrow. Anji Parker just put a thing on Facebook saying what a credit to the sport Sheffield fans were. [1] I told Neil about it. I've just been talking to Shane now. He's gonna have his farewell here. He's first here every week. He's the only rider who's here before I get here at quarter to five every week. On Sunday he rode in the PLRC, went back to Nuneaton, bought some spares for his bike and then drove to Glasgow that night. What a professional!" Jeff stops to ask passing fans, "Did you come Sunday?"
"There weren't much passing."
"It were rain."

Jeff explains the National League venue allocations, "They tossed a coin and we get the semi-final and they [Scunthorpe] get the final. Last night, heat 11 started at 10.15pm!" Jeff knows many fans personally. "Tim [Gardner] comes from Norwich every Thursday! He's one of your longest travelled fans."

[1] "Whilst I'm online I'd also just like to applaud the dedicated group of supporters that follow the lads to away meetings. Whilst I do it because of my relationship with Shane, you guys do it because of your dedication and support of the team. I don't think I've ever seen such a large group of fans so committed to a club. It's a shame the boys haven't performed as well as they had hoped on the road but your support is very much acknowledged and appreciated. Rye House, Sheffield then Glasgow in 3 days?? You guys are nuts!!"


26 Shades of Shale

[Tim] "It's 155 miles here."
[Jeff D] "It's not the miles, it's the roads."
[Tim] "They're mostly single lane. It usually takes three hours. I've been to Rye House, Sheffield, Glasgow and here and tomorrow Plymouth."
[Jeff S] "How many miles is that?"
[Tim] "I haven't kept count, I try not to!"

Photographer Andy Garner still clutches his framed photograph of Shane Parker. Tim Gardner's black-collared top – with its subtle Sheffield speedway logo – catches Neil Machin's eye, "Tim that top looks fucking excellent! Jeff, I just spoke to Tony Jackson and that story that Keith Denham had a heart attack has a flaw – Tony was talking to him at half ten in his office."

Tim enjoyed his Glasgow trip, "The Glasgow fans had sunglasses and the Sheffield fans had umbrellas!" Neil strolls off to the entrance turnstiles to gesticulate wildly. Eric Boocock stops by my table I've borrowed from the downstairs grandstand bar, "You wouldn't have been standing here at two o'clock on Sunday."

[Jeff] "It was drizzling?"

[Eric] "It was raining more like. The track was perfect at eleven for the biggest meeting of the season and it ruined it. No one could race on that! It were balling up! Graham's [Trollope] the only one who can sort a track out then. He dragged it off onto the outside but then it started fluffing up! You couldn't run it on the Monday with all the fixtures left – so it had to go ahead. Glasgow have seven. They've rained off Ipswich four times this season (they hadn't travelled luckily) and Plymouth had five left but have been running sidecars. I think it's brilliant there. I get on with Mike but then I can tell him he's a silly sausage. Did you hear about the Berwick meeting?" Apparently, nowadays, after a crash at Plymouth, they stage some distracting but innovative entertainment around the bar area. Speedway fan Ken Brown isn't overly taken with the title of my latest book, "It's a Bouquet of Flowers not of Shale."
"I can't change it now but what would you suggest is the title of my next book?"
"How about *The Methanol of Speedway*?"
"They're published by Methanol Press."
"Methanol is the highest octane fuel you can get – it's very dear. It were £7 or £8 a gallon before they went to that litre lark. They're only allowed so much in these vans because of the risk of it blowing up. Diesel and methanol together isn't good. There, you've got a bit off me for nought! Morris's make it."
"Morrisons?"
"Morris's not Morrisons!"
"You know a lot about it."
"I've been going to speedway since I was 10. I went to Belle Vue at Hyde Road. Have you heard about that?"
"I have, it's the old track."
"Hyde Road should never have come down, it were built for speedway. They liked to get it finished by nine or quarter past so you could go to the fair afterwards and spend your money!"

GRT Media supremo Bob Tasker suggests I visit hospitality box number 2, "I'm six or eight foot from the best place I've ever stood to watch racing! Come over and see for yourself." Bryn Smith arrives with his daughter. If the (now demolished) Sheffield training track gave many riders their first steps in the sport, according to Bryn something similar applies to Britain's most famous speedway commentator. "Nigel Pearson used to do the centre green here you know, he was very good."

At Sheffield you can tell when the riders are about to come out on parade since the stadium speakers loudly blare into life with a pre-recorded message that wouldn't be out of place in the Monster Mash, WWF or Michael Jackson's Thriller video. We're told we're about to witness something "Live and direct from the darkest depths of your imagination!" Bizarrely, it transpires our collective imagination is speedway oriented and this psychic trip of a lifetime takes place at the "home of the Sheffield Window Centre Tigers!"

Until about August, most neutral observers rated both Sheffield and Workington as likely challengers for the Premier League crown (should Glasgow falter). However, Workington's injury crisis and various silly dropped points for Sheffield have seen both these challenges to the rampant Glasgow Tigers go off the boil. If there were PL Play-offs, then an element of uncertainty would still remain but, as it is, Sheffield look likely to occupy the runners-up position in 2011. If speedway were raced over three heats, tonight Sheffield would win 14-4. Heat 4 sees regular Workington guest Claus Vissing make a jiggly start from gate four before Ricky Ashworth knocks him off with a flamboyant sweep of his back wheel. Surrounded by paramedics, Vissing remains prone on the floor for 30 seconds and then sits on the track for two minutes while the Comets team manager Tony Jackson leisurely strolls over from the pits in his trademark oversized Comets Wulfsport anorak. Tony and Claus then chat animatedly on their walk back to the pits, while Dave Hoggart reveals exciting news for Sheffield Tigers fans' diaries. "There's a dinner and social on November 3rd that hopefully you'll support to thank Shane for all he's done and his year with us! It's a carvery meal and live band right through until midnight. Its £15 admission – which is cost price and the rest we'll chuck into Shane's fund – and we'll work something out for the kiddiewinks. It's your chance to thank Shane for everything he's done here and for speedway! The band will be the Magoos. If you went to Cardiff, you'll have seen them. They went down a storm and this is the first opportunity I've had to bring them back into the public eye since then."

Referee Craig Ackroyd gives roughly five minutes – to allow for repairs to Vissing's bike – before the Comets romp to a maximum heat advantage. Jeff Dooley surveys the scene from his vantage point at the trackshop, "It's not a very big crowd tonight. We didn't have a meeting last Thursday but with Sunday – when it was £20 for the PLRC – it's asking a bit much to get two meetings out of one pay packet! Workington would be exciting, if there were Play-offs, but all we've got left is the National League semi-final second leg next week and the Americans the week after so it's going to fizzle out, I'm afraid!" Heat 5 sees Workington hammer in another maximum, while Vissing wins heat 6 with an outside pass around Shane Parker. Prior to heat 7, news from the pits compounds perceptions that Sheffield might struggle. "Unfortunately because of that arm injury on Sunday, Ricky Ashworth has withdrawn from the meeting." With three programmed rides to go, let alone his likely nomination for heat 15, Ashworth's withdrawal is suboptimal. Though Simon Lambert overtakes Simon Nielsen to secure second he's unable to make any impression upon the race leader Richard Lawson and, with Birks last, this Comets heat advantage ties the score. Craig Ackroyd's decision to declare heat 8 an unsatisfactory start prompts loud grumbles from the Sheffield fans who I suspect base their knowledge on the rulebook Kelvin Tatum uses for his Sky Sports Monday night live broadcasts. Predictably enough, nowhere first time out, the Comets secure a heat advantage (Vissing wins again) to lead by 2 points. It's a notable turnaround, especially given they trailed by 10 after three heats. During the track grade prior to heat 9, David Hoggart banters with the crowd about the unseasonably warm weather. "Now then, if this was Glasgow, we'd sun this off on a nice day like this!" Though Kenny Ingalls wins heat 9, the real action sees Emiliano Sanchez monster Claus Vissing to snatch second on the line. For the last two laps of heat 10, Richard Lawson conducts a masterclass in intelligent pressure and his diligent search for a gap comes good when he blasts round to beat Hugh Skidmore by the proverbial tyre knobble.

Just before Shaun Leigh reveals the line-up for heat 11, David Hoggart interjects from the centre green, "I think you'll find we're having the interval."
"Oh, are we?"

"You don't listen?"
"Err, no, okay."
"Here at Owlerton we have the best track staff – so everyone says who praises them! Each year the track staff have a vote amongst themselves for the Rider of the Year so it will be absolutely no surprise for you to learn that Shane Parker is the rider!" Shane strides onto the centre green to receive the framed photograph of himself brought along by Andy Garner. David Hoggart tees-up the conversation by noting 2011 is a "good way for Shane to go out with so many points and a good year at Owlerton."
[Shane] "I'm not enjoying tonight but that's another story. I've thoroughly enjoyed it here and wish it was more than one season but there's young kids coming through the sport and me body's killing me, I'm getting older! To be honest, I wish I'd come here earlier."

Talk of the 2011 Premier League season inevitably turns to the topic of the champions, Glasgow. Shane praises them, "They deserve to win to be fair. Unfortunately, it's not one of the years I was with them. Unfortunately I've not won a championship in my career!" Shane continues, "I enjoy the racing here and some of the other tracks are dire" before looking ahead to his social in the Panorama Room, "I feel very honoured to be part of it. I'm sure the family will come. I know I'll enjoy it! I'd just like to thank all the supporters and all the people who paid to come in, we wouldn't be able to do it without you."

A speedway fan with two young daughters lingers by my table. "I've had 20 years away in the Army. I used to go to Long Eaton. For 15 years, I couldn't tell you if the tracks were there or not! Then, I went to Berwick. This won't be here in 10 years unless they get the youngsters in!" Ex-Weymouth Wildcats photographer and press officer Julie Martin stops by with her boyfriend Mark 'Buzz' Burrows but can't chat because the Panorama Room beckons, "I have to go because I haven't had my dinner!" Like many speedway fans, Jeff Dooley keenly anticipates the new national speedway stadium coming on stream in Manchester. "Belle Vue's a funny place, they seem to have blinkers on and only look 100 yards into the future. They've been looking at stadiums everywhere. They were in Chesterfield last week looking at that one." A Sheffield fan stops at the trackshop with questions that require no answers. "How come Topinka isn't here? And how come they have a guest? Ricky's arm and shoulder aren't good. He can't really hold onto the bike – why didn't he tell them on Sunday to get a guest for Glasgow and tonight? Auty wants to go Elite League but, come on, not at the moment – he's a good rider but not yet there! I wonder if Joe Screen will be here next year?"

Ricky Ashworth's withdrawal from the meeting sees Ashley Birks take his place at the start line for heat 11. Ashley leads the race until passed in quick succession by race winner Rusty Harrison and his Comets teammate Kenny Ingalls. By the end of the second lap Ashley is fourth behind Simon Lambert. The Comets third maximum heat advantage isn't taken in good heart by a small section of fans away to my left who wind themselves up with cries of "What a load of rubbish" and "Get a grip!" Workington's amazing turnaround from 10 points down after three races now sees them 8 points up after eleven!

All the real passes in heat 12 happen on the last lap. In the battle of the Simons, Lambert overtakes Nielsen on bend one while Auty passes Ingalls down the back straight. Temporary teammates for the night, Harrison and Vissing then conduct a masterclass in how not to team ride. On the first corner of heat 13, Rusty rears dramatically to slow Vissing enough to enable Shane Parker to storm round the outside. To my real-time untutored eyes, on the third bend the yellow-helmeted Vissing appears to pull a slight locker on the outside line that causes the blue-helmeted Ashley Birks to clatter into his back wheel and, thereby, splatter both riders and their bikes into the safety fence in a dramatic fashion that borders on the horrific. It's precisely the sort of incident that Sky Sports endlessly replay. Almost as one, the home-straight grandstand crowd stare through the downstairs bar windows to watch the race replay on the television monitors. Given the real-time ferocity of the crash to the naked eye, I'm surprised such replays get broadcast. Repeat viewings only mostly confirm previously held initial prejudices amongst a boisterous home crowd. Seen live, I would disqualify Birks for clattering into Vissing's back wheel but, when seen on the replay, I'd hold Vissing responsible for leaving nowhere to go but into his back wheel. Jeff Dooley's in no doubt, "He locked up and left Ashley with nowhere to go. Ashley's young. A few years back, Paul Cooper threw his helmet at Vissing as he went past. It missed by a million miles and then they had a fight!"

During the break for fence repairs, bike repairs as well as the delay for an ambulance to arrive to take Vissing to hospital, kindly softly spoken bearded Sheffield Tigers fan David stops by to chat. He takes a straightforward approach to his speedway viewing pleasure. "I just like to see young riders improving! I don't bother with the gossip. It'll take them all night to rebuild the fence. Whenever Vissing comes here something happens! A few years ago – I can't remember it exactly – he took Paul Cooper to the fence and Cooper waited by the pits gate but Vissing went wheeling past waving to the crowd and, then, they had an argument. If it's not that, he's always on the shale!" Shaun Leigh updates the crowd, "I can tell you that Claus Vissing will be taking a trip to the hospital. Once the medical staff are free again and the fence rebuilt, we'll get on with it." David listens intently, "I was surprised they showed it again on the stadium monitors because they looked badly injured!" Apart from their now injured guest Vissing, tonight Comets resistance comes from Harrison, Ingalls and Lawson. "Lawson's doing well."
"Just like his dad."
"He never moved up and was happy to dominate in the National League."

While the wait drifts on, it's still warm enough to want to hunt out the ice cream van. Some fans give up and stream out. Clutching possibly vital paperwork, Andrew from the Speedway Office wanders determinedly against the tide, "That hasn't done much for Vissing or the fence."
"Why are people leaving?"
"Well, its quarter to ten, the ambulance has to go to the hospital and come back again and it's close to curfew at ten. We can go five or ten minutes past but no longer so there might be no more racing." Almost as soon as the words leave his mouth, the rerun of heat 13 gets underway. It's a three-rider race won by Shane Parker. After David returns to collect his spectacles, "I thought I couldn't see heat 13!"

While we wait for the penultimate race, David Hoggart reminds the crowd many items are available at bargain prices at the trackshop. "Jeff tells me, that their [replica] shirts were £32 and they've dropped them to £25!" Jeff's proud of the merchandise, "I had to design it myself."
"It's nice to see the Window Centre on the back."
"I wanted to do that for him because he's a good friend of mine and a good friend of Sheffield speedway."
"Was he pleased?"
"Chuffed to bits I was going to put the name there but the designers put the logo on, after taking it off the website. This season the riders have to wear the shirt, cap, or jacket before and when they walk round – otherwise they get fined and it goes to the Ben Fund."
"Let's hope you don't change sponsors!"

Lawson leads heat 14 until both Sanchez and Lambert storm past for a Tigers 5-1 that ties the scores (42-42) with one race to go. It promises to be quite a finale. Though it's more or less on the ten o'clock curfew, Dave Hoggart advises the crowd, "I've just had a word with the team managers and we're going to run some second-half races because they've been ready over there for 30 minutes and we want to give them a skid! If you come from Workington, I apologise because there won't be a parade if you win but neither will there be for Sheffield!"

With pressure for points and a swift start, there's visible movement on the outside gates. I confidently tell Jeff, "Yellow", he nonchalantly nods towards the lights, "Blue – Auty!" Already on the outside gate, in the rerun Auty begins with an additional 15-metre handicap. On a fast track like Sheffield, you're unlikely to make up such a distance. For the first two laps it looks like we're definitely in for a drawn heat. However, Josh noticeably gains greater momentum on laps three and four to somehow snatch third on the line from a gobsmacked Richard Lawson who – like many Comets fans in the crowd – throws his head back in disgust! After noting the fickleness of the fans ("Seven heats ago they were booing them and now they're cheering!"), Jeff Dooley ruminates on how track conditions impact the home riders, "On Sunday it were deep and caught them all out and they probably came tonight expecting the same and it's as slick as Rye House – bare as a baby's bottom!" I can't resist a word with the Sheffield fans who'd earlier questioned Josh Auty's performance levels, "Your favourite rider won it." "Okay Auty did but it could have gone either way could that!"

Though the line-ups don't appear in the programme, remaining fans get to see some second-half races too. Afterwards Craig Ackroyd stops en route to the car park to discuss events with Sheffield's David Hoggart who's not totally enamoured with Claus Vissing. "He's been doing that all season – turning left! He's a good rider but that's his reputation. Many of the foreign riders do it to stop you getting past. He turned on Emiliano but he just carried on or he'd have been like Ashley. When they got back to the pits, we knew they'd have words but Emiliano just went up and hit him! Can you imagine that? Emiliano? Vissing said, 'I will choose one of you' and Ashley was the one he chose. You just don't do that, let alone to a rider like Ash. It would be like Ivan Mauger doing that to a second halfer. Ash is only young. He had nowhere to go. He's hurt his hand and his bike is wrecked. The fence posts snapped. If you turn left, it's only a matter of time before that happens."

Up in the bar, people discuss the Vissing/Birks crash. Julie Martin tells Buzz Burrows, "I just saw one rider go into the back of the other." With his vast experience, Buzz has seen (and felt) it all before. "If you lock up, they've got nowhere to go and it's your fault."
[Julie] "That's why it's good you're not riding anymore."
[Buzz under breath] "Humph."
[Julie] "That's why I'm not a ref, I get it wrong every time! You were a rider so you can see what's what."
[Buzz] "Ninety-five percent of the ref's haven't ridden."
[Julie] "Why don't you become a ref?"
[Buzz] "Dave Dowling wrote to me asking if I'd like to!"

**29th September Sheffield v Workington (Premier League) 46-44**

# Afterword

I hope that you enjoyed the journey.

Phew! What a fantastic year I had and what memorable experiences. I was genuinely overwhelmed with the kindness of strangers and amazed how people went out of their way to help me! Thank you so much! There are, of course, exceptions to every rule.

Obviously all mistakes remain my own. If you have any comments, of either persuasion, please get in touch via my website on www.methanolpress.com I would be delighted to hear from any copyright holder to make the appropriate credits or acknowledgements.

I mentioned earlier that I have been overwhelmed with help and kindness. I hesitate to name everyone as, inevitably, I will make a mistake and miss someone I'm extremely grateful to, so, with sincere apologies to those who I do manage to miss out I would like to thank the following people: Peter Adams, Rachael Adams, Mike Amos, Katherine Anne Anne, Graham Arnold, Ian Attwood, Mike Bacon, Pete Ballinger, Robert Bamford, Nick, Johnny, Bev, Molly and Colin Barber, George and Linda Barclay, Dick Barrie, Ian Belcher, Malcolm Blythe, Alan Boniface, Tim Booler, Mick Bratley, Jim and Steven Brykajlo, Paul Burbidge, Brian Burford, Peter Butcher, John Campbell, Alison Chalmers, Jonathan, Keith and Cheryl Chapman, Steve Chilton, Jon Cook, Graham Cooke, Dougie Copland, Mike Corby, ICA Crook, Gordie Day, Alan Dick, Steve and Debbie Dixon, Ann and Jeff Dooley, Chris Durno, Graham and Denise Drury, Neil Dyson, Kay Ellis, Jim Elsdon, George and Joan English, Dave Fairbrother, Jim Fleming, Richard Frost, Chris Gay, Arnie Gibbons, Bill Gimbeth, Steve and Tai Girdwood, Rob Godfrey, Mike Golding, David Gordon, Rob Greig, Andy Griggs, Patricia, Jonathan and Tim Hamblin, Debbie Hancock, Steve and Jan Harland, Brian and Marge Havelock, John Hyam, John, Jordan, Karen, Mark and Judy Hazelden, Richard Hollingsworth, Dave Hoggart, Mike Hunter, Tony Jackson, Sue Jackson-Scott, Billy Jenkins, Sally Knight, Sheila Le-Sage, Kevin Ling, Kevin Long, Gary Lough, Chris and John Louis, Michael Max, Ian, Alan and Jean Maclean, Neil Machin, Julie Martin, Keith and Mandy Mason, Dennis McCleary, Keith McGhie, Charles McKay, Allan Melville, Steve and Sarah Miles, Chris Morton, Mike Moseley, Chris Moss, Jayne and Joshua Moss, Hazal Naylor, Martin Neal, Bill Norris, Peter Oakes, Paul Oughton, Gordon Pairman, Nigel Pearson, Rob Peasley, Di Phillips, Gary Pinchin, Mark Poulton, Andy Povey, Colin Pratt, Ian Presslie, Dave Rattenberry, Steve Ribbons, Dave and Margaret Rice, John Rich, Gareth Rogers, Laurence Rogers, David Rowe, Wayne Russell, Craig Saul, Mark Sawbridge, Rob Scrutton, Len Silver, Andrew Skeels, Tony Steele, Barry Stephenson, Trevor Swales, Graham Tagg, Bob Tasker, Peter Toogood, Stuart Towner, Christina Turnbull, Chris Van Stratten, Nick Ward, Dave Watters, The Reverend Michael Whawell, Bryn Williams and Malcolm Wright.

I also owe so many 'thank yous'. The book wouldn't look as lovely as it does without Vicky Holtham. The look and feel of the covers, website and book layout were all originally conceived by Rachael Adams. There would be many more errors than there are without the help of Vy Shepherd. Graham Russel has shown tremendous pedantry, patience and knowledge to wrangle with my words to convert them into some sort of sense. Michael Payne was, like many others, there at the start and Billy Jenkins continues to offer encouragement, advice and support. My true friend Sue Young encourages me often in so many things. Of course, without the love and guidance of my parents – Mary and Alan – none of this book or so many other things would ever have been possible. Finally, you can never have too many teachers and I was lucky enough to have been inspired to write my speedway books by a truly great teacher, poet, musician and wit – Michael Donaghy. He remains greatly missed.

Before I go, we only get one chance to make a first impression – this holds for speedway clubs as much as people. If you've read to here, it's safe to say that you've probably been bitten by the speedway bug (for which, apparently, there is no permanent antidote). We can all no longer go to speedway for the first time. However, we can take new people and relive this experience vicariously once more, albeit second hand. I always like taking newcomers.

It just remains to say: if you go to speedway already why not make a point of taking even more friends and/or first timers this year? And, if you haven't been for a while or have never been, now is as good a time as any to start!

Finally, over these past few years it has often been a joy and privilege to meet so many lovely "speedway people" and also to have the chance to write about British Speedway.

Yours in speedway!

Thank you for coming on my last journey round the tracks.

Brighton

20 August 2012

# ALSO AVAILABLE FROM METHANOL PRESS

## *"just a saddo with no life"* MIKE BENNETT

### Bouquet of Shale

*"works fantastically well…an engaging writer with an incredible eye for detail"*
*Speedway Plus*

274 pages 70 Black & white photographs
**£20.00** Paperback

### Shale Trek

*"Scott's look at the world of speedways has few equals. Why? Because he tells it as it is. He speaks the fnas language"*
MIKE BACON, *Ipswich*

*"Scott's look at the world of speedways has few equals. Why? Because he tells it as it is. He speaks the fnas language"* MIKE BACON, *Ipswich Evening Star*

*"a different perspective on the weird and wonderful goings-on … fantastic … British speedway's voice of the people refuses to be silenced"*
PAUL BURBIDGE, *Speedway Star*

306 pages 120 Black & white photographs **£20.00** Paperback

### Quantum of Shale
#### Around the Tracks

*"a prose that possesses a kind of petrol-driven Dirty Realism, as though Raymond Carver had decided to turn up at a speedway meeting"*
IAN MCMILLAN, *The Times*

328 pages 190 Black & white photographs **£20.00** Paperback

### When Eagles Dared

*"A remarkable read and something completely different to any other books on the overcrowded market. Jeff has a very personal style that you might either love or hate but I found it riveting and enjoyed almost every word"*
PETER OAKES, *Speedway Star*

172 pages 158 Black & white photographs **£10.00** Paperback

### Concrete for Breakfast
#### More Tales from the Shale

NOMINATED FOR BRITISH SPORTS BOOK OF THE YEAR AWARD 2009

*"Jeff Scott's* Concrete for Breakfast *is possibly Speedway's War and Peace, or its Ulysses, or, in some of the chapters, its local newspaper gossip page. It's the epic by which all other books on the sport will be judged … Read this book for a glimpse of a lost tribe, for an examination of collectivism and individuality somehow working together, for endless descriptions of English weather, and for a brave attempt to pinpoint a particular branch of human endeavour that often seems to be far from the centre of things."*
IAN MCMILLAN, *The Times*

302 pages 223 Black & white photographs **£20.00** Paperback

### Showered in Shale
#### One Man's Circuitous Journey Throughout the Country in Pursuit of An Obsession – British Speedway

*"It's all here, in almost soap opera style proportions … this is British Speedway stripped bare, this is how it really is and some of it isn't pretty "*
BRIAN BURFORD, *Speedway Star*

508 pages 409 Black & white photographs **£20.00** Paperback

### Shifting Shale
#### 2006 A Race Odyssey

*"As always the beauty of his [Jeff's] work is in the description of the minutiae"*
JIM HENRY, *Speedway-Researcher*

345 pages 143 Black & white photographs **£10.00** Paperback

### Shale Britannia
#### A Sideways Glance at Speedway

*"A marvellously evocative book"*
JOHN INVERDALE, *Daily Telegraph*

256 pages 345 colour photographs **£10.00** flexiback

All available from **www.methanolpress.com**

methanol press